EPIC THOUGHTS: THE BEST OF

By Austin P. Torney

Copyright 2015 Austin P. Torney

ON THE ORIGIN,
WHO DESIRED THAT ON ITS TOMB
SHOULD BE INSCRIBED—

"Here lieth One whose name was writ on water."
—Shelley

The 'false' and melted vacuum was liquid energy—
Unstructured, unordered, and going nowhere,
But then, inexplicably, it 'fell',
As from a kind of 'shelf',
Whirling, twirling, and swirling inward
Until there was no more inward left...

It 'thought' that its future could never be,
That its quality was but written
On the water and the wind
With a feathery quill
Whose ink was the smoke and fog
Of a shimmering dream.

Then it died... like the Phoenix;
And thus it crystalized, frozen,
Into our structured 'true' vacuum...

For, ere the breath that could erase it blew,
Death, in remorse for that fell slaughter,
Death, the immortalizing winter, flew
Athwart the flowing stream—
And Time's printless torrent grew
A scroll of crystal,
Blazoning the name
Of
'The Universe'!

(Shelley altered)

AFTER THE STARS HAVE GONE— THE FINAL, SILENT DARK

THE LAST CHANCE SALOON (CASINO)

Entropy is always the winner in the end,
When there's no more money left to lend;
Meanwhile, we stabilize, in nature's way,
Rearranging resources temporarily.

Prelude

Going beyond our very old obsession so vast,
Of how it all began, back in the distant past,
Yet retaining our search for meaning, from that,
We now turn to how will it all end, this and that,
Whether becoming collapsed, expended, or flat.

Is there is some deep meaning in all that?

Yes, for it is there in that future distance,
We'll find or not the end of our persistence,
Whether or not we are at all forever resistant,
Whether all that was and what was did and done
Will be of any long-lasting benefit to anyone—
Of what destiny awaits, if there ever was one.

Endings are important to us, as what we're about,
Because we believe that how things turn out
Implies what the beginnings ultimately meant,
Of what or not is our place in the firmament.

As an ambitious species of nurture and nature
We now and have always pointed toward the future,
For of the three forms of the chimpanzee:
The common chimp, the bonobo, and us, we
Are the only chimp who went beyond the trees...

And more importantly, ever out of Africa freed,
By that exodus, which laid down, indeed,
From that experience, the urge and the need
To move on, exploring, ever planting another seed.

The horizons on Earth sufficed us through time
For many millennia but now the horizons' climes

Have broadened through cosmology and physics,
And so they can well inform us of our prospects.

The future matters to us for very basic reasons:
We wish to offset our mortality, our pleasin's,
To know if humanity's works for every season
Will be remembered, or lost—all for nothing, even.

The Final, Silent Dark Marches On…

*Time hurls a million waves of its displacement
At us, yet we are still here—the replacements.*

*Time ever gray with age hurls its changes then,
'Gainst existence's rock, time and time again,
The entropic seas denuding the sands,
Yet energy is preserved via science's wands.*

Reminiscence has weathered but could ne'er wither,
For in the mists of time yesteryear yet appeared.

*Would the prospect of a "Big Crunch" bring on phobia,
Such as an ever more confining claustrophobia?*

Seems a better thought, somehow, though no picnic,
But more pleasing if the universe were to be cyclic,
Although then all would still be really crushed,
And forever lost, gone headlong into the rush.

We expect cycles, for all the days and seasons
Embedded this in our ancestors, into our reasons,
Since at least the periodic supplies some rhythm,
A pattern—the rolling hills of lives onward driven.

As for the cyclic, endless repetitions, they too
Would seem to revolt more of us than just a few;
As too perhaps would some infinite abyss of time,
Which both grant us neither reason nor rhyme.

*Does the drama go on forever, or does it end?
What do the visions of the future portend?*

*Doesn't it all have some purpose meant—
A goodly end that all of it to us might it present?*

Is our higher mammal time certainly

But of such a short parentheses within eternity?

It's only a finite time then, which too tends
To horrify so many, as the universe ends,
Such as told by Robert Frost, a name of chill:
In heat or in cold, known as fire or ice, still.

Should we not believe in God since nothing lasts?

Well, if nothing lasts then of what our purpose past?

Is a purpose really required, so constructive,
Or would that really be quite restrictive?

No realm could really be special or sent,
Its becoming being of some specific intent,
For all has arrived as the causeless non-precedent.

Is there anything wrong with the freedom to be,
Anywhere, any how, or any time during eternity?

Should we rail against the law of entropy—
The "heat death" of thermodynamic energy,
The second of its final laws, you see,
Because it would destroy all of history?

There are so many ways for disorder to be
Than any one ordered state specifically.

Would even a heaven on Earth become a misery
If as it might, contain no more novelty?

Must there be an end to our revelry?

Can't we at least hibernate eternally?
Won't all matter too last eternally?

Will Shakespeare's works live on, paternally?

Is this not a Wagnerian struggle for eternity?

Science Can Settle Whether a Last Day
Is Ever Going to Come this Way

Only a decade or so ago, with consternation,
We discovered the universe's acceleration,

Its expansion even increasing, onto a thin disaster,
The galaxies getting further away ever and ever faster—
Then one last snapshot taken, for all to remember.

The accelerating expansion of the universe's rafters
Means that the universe will cool even ever faster,
So any rare forms of the future's life prolongers
Will have to keep themselves ever more cooler,
Think more slowly, and hibernate ever longer.

One day the protons will even fade away,
Leaving but dark matter, electrons, and positrons.

THE WAVES OF THE ANCIENT SWELLS
OF TIME'S FORGETTING SWELLS
SWEPT EVER ON...

As Time, now hoary with age,
Hurled forth its ashen change,
The charge ever san, pale and colorless,
That force born to summon decay, so endless,
'Gainst Nature's Universe, each and every day.

Time and time again Time fed all upon,
In its bloodless, white, and waxen way,
But this everlasting rose would not fade,
Its luster even brightening by the day,
Ever unsuccumbing to the sickly, peakèd
State draining drawn the life away.

Entropic seas yet denude the mountains,
Yet this enduring flower never-endingly
Has cast Deathly Time aside, fas now,
Ceaselessly somehow thriving on
To that which is the near imperishable,

The flame of beauty still inextinguishable,
Forever celebrated as immutable,
Gaining a seemingly perpetual permanence
From the undying love of the glorious truth.

Yet everything was moving apart, cooling off,
The big slowdown not really so very far off;
Ultimately, even the black holes of late
And the lightless planets will dissipate.

The primordial soup once so rich and hearty
Was now a thin gruel that couldn't serve the party.

One day every particle will be moving away
From every other particle, so much out of the way
That they won't even be able to see one another;
Thus for all intents motion will have ceased forever.

Our spurt of life followed by an infinite stretch
Of dark equilibrium was but the briefest sketch—
A warm and fuzzy stage, so interestingly active,
Whose time relatively was but infinitesimive.

Yet we were there in all our glory,
For whenever else could we have been?

In the future, uncounted societies of
Overlapping minds accumulate, with love,
In island redoubts, their preserved data burning
With a vital remembrance, in which, returning,
The past is the present and future, they all reliving
The data, even animating it, and ever altering.

Without any new enrichments the present and future
Reprise the past in this retreat from external nature.
Their candles would have been nearly invisible to us—
They enduring by diminishing so as not to exhaust.

They made few new memories, a kind of blind sight,
For whatever realities had ever existed out of sight
Of their own mental structures were now fractured,
And thus not so different from those manufactured.

The Penultimate Part of the Final Dark

AN ESCALATING ONE-WAY TRIP
FROM A FLUKE TO OBLIVION

*The majority of the energy
Of the universe is dark today,
Although everything else passes
Through it in every way.*

*It's everywhere,
Having a component
That repels its own state,*

Which cause the expansion of
The universe to much accelerate.

DARK ENERGY MATTERS: THE ESCALATION

We're on a one way trip from a quantum fluke,
That maximal energy within old Planck's nook,
Heading toward the oblivion of sparse expansion—
All that we ever loved and knew going to extinction.

They sent message of early warnings to some,
In those castles of illusion, yes, many a one—
That they would face the decay not so far away
Of the heavy particles—the "proton pause", one day.

No self-assembled granularity can endure
Forever but must return to the substructure,
And so the lives must all transition, it seems,
From heavier to much lighter regimes...

Although this too would not be permanent—
All destined to be swallowed by the firmament.

We have often asked why some space exists,
Why it permits the countless to briefly persist
On Mother Earth, nourished under Father Sky—
All of those finite sparks that light and die.

There were those who endlessly debated
Whether to live in their virtuals unabated
Or to press forwards and outwards, of delirium,

To seek out new localities in the mysterium,
But the pauses of the heavy particles continued,
And so there was nowhere to go for the retinued.

It was much simpler once in those days of old
When we thought that universes didn't go cold,
But that they expanded and then collapsed,
Still destroying all, yet ever giving more to last.

And well before that, once upon a storied time,
We simply made it all up, with tales and rhyme,
In place of any physical observations
Such as through revealing experimentations.

The past was now a reef of dead accumulation,
A graveyard of various useless information,
Which despite its splendorous beauty
Could not provide for a novel futurity.

The last one of us, born of the sparkness,
Kept a window to the outer darkness...

She looked out from a once brightly
Colored and sparkling inner reality
Into the dark abyss...

There was nothing out there,
All being so lonely and bare—
No more singing of life's song,
For now everything was gone.

The Final Epilog

There could not have been any specific time,
One that was privileged over any other chime,
Nor any special place, nor any certain form
Arising out of the necessarily causeless realm.

Even the locally specific dates and places past
Of the events' novel memoirs couldn't last,
They being writ on water, with no meaning vast,
Disappearing in significance so very fast,
Since it's only the universals that last.

The protons were now gone from the show,
Having decayed so very long ago
Into positrons—ever canceling the electrons,

And emitting the fleeing light of photons,
There being of course an equal amount
Of protons and electrons in the count.

And of course along with all the protons
Went all of the atomic elements—the end,
All of their forms becoming myth and legend,
As they were still dreamt in night dreams,
Those forms that we once had, so it seemed.

She, as many of a luckily adaptable kind,
Had long since lightened and lighted her mind,

With the dwindling electrons and precious photons—
That beginning light of ancient times growing wan.

Ours had been the only line in the universe,
One that had become sentient, with proto-man first,
The rest of the cosmos being but a colossal waste,
A foreboding, harsh, and very dangerous place.

She was now the only one left,
Having outlived all of the rest.

The universe was near crumbling away,
Having run out of space, time, and all its sway.

She was dispersing, melting, into the vacuum, lone,
But she held on for another thousand years, alone,
And then she too was gone,
Being the last of the hominid's song,

Of all that was sapient: the *Magnificat*,
The composition of Earth's sweet plot,
The greatest symphony that was ever sown,
It now having faded into the unknown.

From near nothingness our forms became,
And into the same must go the remains.

If the unknown be such, though it's otherwise;
But if it's yet called 'unknown' then the reply
Is still for sure that we're free to be, anywise.

If you've shed a tear reading here
For both the far and the near and dear
It won't make their graves green again,
But it's possible that life could begin again...

Be of Good Cheer—the sullen Month will die,
And a young Moon requite us by and by:
Look how the Old one meagre, bent, and wan
With Age and Fast, is fainting from the Sky!
(A Khayyam quatrain that's not in FitzGerald's Rubaiyat)

Our fruits are of a universal seed
As the yield of All possibility treed,
And siblings elsewhere in the entropic sea
Will also be born of such probability.

The Eternal Return

Behind the Veil, being that which ev'r thrives,
The Eternal IS has ever been alive,
For that which hath no onset cannot die,
Nor a point from which to impart its Why.

Some time it needed to learn Everything for,
And now well knows how these bubbles to pour,
Of existence, in some like universe,
As those that wrote your poem and mine, every verse.

So as thus thou lives on yester's credit line
In nowhere's midst, now in this life of thine,
As of its bowl our cup of brew was mixed
Into this state of being that's called "mine".

Yet worry you that this Cosmos is the last,
That the likes of us will become the past,
Space wondering whither whence we went
After the last of us her life has spent?

The Eternal Saki has thus formed
Trillions of baubles like ours, and will form,
Forevermore—the comings and passings
Of which it ever emits to immerse
Of those universal bubbles blown and burst.

So fear not that a debit close your
Account and mine, knowing the like no more;
The Eternal Cycle from its pot has pour'd
Zillions of bubbles like ours, and will pour.

When You and I behind the cloak are past
But the long while the next universe shall last,
Which of one's approach and departure the All grasps
As might the sea's self heed a pebble-cast.

SEASONINGS

Nature springs from winter's tomb,
The bloom already in the seed,
The trees within the acorns.

Surging sprigs sprout from the soil;
Spring showers make the summer flower.

Summer wakes from spring's dying kiss,
Blooming when the rose does,
Sunning after the spring's running.

Summer reigns upon the land,
Eventually fading in the night.

Autumn falls as summer leaves,
Harvesting its sum of days,
Seconding the rose of spring.

The smile meets the tear—
Fall's embers last through December.

Ice winds stalk the weed flowers,
The ghosts frosting the dead stalks,
Snow crystals barring all that grows.

Winter is life cooled over;
Melting snows feed spring waters.

WHO AM I?

I am the brain; I have many layers.
I control thoughts, feelings, and the body,
From inputs within and without,
Interpreting and re-presenting them
With a better and more useful face
Painted upon the raw waves and frequencies.

My senses are as spy outposts upon reality;
Inside, my own developed language of qualia
Portrays the results of the messages produced
By my vast array of neural connections—
Of their analysis, which takes a bit of time.

ALL THAT LIES BETWEEN

Energy is a beauty and a brilliance
Flashing up in its destructance,
For everything isn't here to stay its "best";
It's merely there to die in its sublimeness.

Like slow fires making their brands it breeds,
Yet ever consumes and moves on, as more it feeds,
Then spreads forth anew, this unpurposed dispersion,
An inexorable emergence with little reversion,

Ever becoming of its glorious excursions,
Through the change that patient time restrains,
While feasting upon the glorious decayed remains
In its progressive march through losses for gains.

We have oft described the causeless—
That which was always never the less,
As well as the beginnings of our quest,
And too have detailed in the rarest of glimpses
The slowing end of all of "forever's" chances.

So then we must now turn our attention keen
To all of the action that exists in-between—
All that's going on and has gone before,
Out to the furthest reaches, ever-more,

For everything that ever happens,
Including life and all our questions—
Meaning every single event ever gone on
Of both the animate and the non,
Is but from a single theme played upon.

This then is of the simplest analysis of all,
For it heeds mainly just one call—
That of the second law's dispersion,
The means for each and every occasion,
From the closest to the farthest range—
That which makes anything change.

These changes range from the simple,
Such as a bouncing ball resting still
Or ice melting that gives up its chill,
To the more complex, such as digestion,
Growth, death, and even reproduction.

There is excessively subtle change as well,
Such as the formations of opinions tell
And the creation or rejections of the will.

And yet all these kinds of changes of course
Still become of one simple, common source,
Which is the underlying collapse into chaos—
The destiny of energy's unmotivated non-purpose.

All that appears to us to be motive and purpose
Is in fact ultimately motiveless, without purpose.
Even aspirations and their achievement's ways
Have fed on and come about through the decay.

The deepest structure of change is but decay,
Although it's not the quantity of energy's say
That causes decay, but the *quality*, for it strays.

Energy that is localized is potent to effect change,
And in the course of causing change it ranges,
Spreading and becoming chaotically distributed,
Losing its *quality* but never of its quantity rid.

The key to all this, as we will see,
Is that it goes though stages wee,
And so it doesn't disperse all at once,
As might one's paycheck inside of a month.

This harnessed decay results not only for
Civilizations but for all the events going fore
In the world and the universe beyond,
It accounting for all discernible change
Of all that ever gets so rearranged,
For the *quality* of all this energy kinged
Declines, the universe unwinding, as a spring.

Chaos may temporarily recede,
Quality building up for a need,
As when cathedrals are built and formed,
And when symphonies are performed,
But these are but local deceits
Born of our own conceits,
For deeper in the world of kinds
The spring inescapably unwinds,
Driving its energy away—
As All is being driven by decay.

The *quality* of energy meant
Is of its dispersal's extent.
When it is totally precipitate,
It destroys, but when it's gait
Is geared through chains of events
It can produce civilization's tenants.

Ultimately, energy naturally,
Spontaneously, and chaotically
Disperses, causing change, irreversibly.

Think of a group of atoms jostling,
At first as a vigorous motion happening
In some corner of the atomic crowd;
They hand on their energy, loud,
Inducing close neighbors to jostle too,
And soon the jostling disperses too—
The irreversible change but the potion
Of the 'random', motiveless motion.

And such does hot metal cool, as atoms swirl,
There being so many atoms in the world
Outside it than in the block metal itself
That entropy's statisticals average themselves.

The illusions of purpose lead us to think
That there are reasons, of some motive link,
Why one change occurs and not another,
And even that there are reasons that cover
Specific changes in locations of energy,
The energy choosing to go there, intentionally,

Such as a purpose for a change in structure,
This being as such as the opening of a flower,
Yet this should not be confused with energy
Achieving to be there in that specific bower,

Since at root, of all the power,
Even that of the root of the flower,
That there is the degradation by dispersal,
This being mostly non reversible and universal.

The energy is always still spreading thencely,
Even as some temporarily located density—
An illusion of specific change
In some region rearranged,

But actually it's just lingering there, discovering,
Until new opportunities arise for exploring,
The consequences but of 'random' opportunity,
Beneath which, purpose still vanishes entirely.

Events are the manifestations
Of overriding probability's instantiations—
Of all of the events of nature, of every sod,
From the bouncing ball to conceptions of gods,
Of even free will, evolution, and all ambition,
For they're of our simple idea's elaborations,
Although for the latter stated there
And such for that as warfare
Their intrinsic simplicity
Is buried more deeply.

And yet though sometimes concealed away,
The spring of all creation is just decay,
The consequence and instruction
Of the natural tendency to corruption.

Love or war become as factions
Through the agency of chemical reactions,
The actions being the chains of reactions,
Whether thinking, doing, or rapt in attention,
For all that happens is of chemical reaction.

At its most rudimentary bottom,
Chemical reactions are rearrangements of atoms,
These being species of molecules
That with perhaps additions and deletions
Then go on to constitute another one, by fate,
Although they sometimes only change shape,
But too can be consumed and torn apart,
Either as a whole or in part, so cruel,
As a source of atoms for another molecule.
Molecules have neither motive nor purpose to act,
Neither an inclination to go on to react
Nor any urge to remain unreacted,
So then why do reactions occur if unacted?

Molecules are but loosely structured
And so they can be easily ruptured,
For reactions may occur if the process energy norm
Is degraded into a more dispersed and chaotic form,
And so as they usually are constantly subject

To the tendency to lose energy, as the abject
Jostling carries it away to the surroundings,
Reactions being misadventure's transformations,
It then being that some transient arrangements
May suddenly be "frozen" into "permanences"
As the energy leaps away to other experiences.

So molecules are a stage in which the play goes on,
But not so fast that the forms cannot seize upon;
But really, why do molecules have such fragility,
For if their atoms were as tightly bound as nuclei,
Then the universe would have died, being frozen,
Long before the awakening of the forms "chosen",

Or if molecules were as totally free to react
Every single time they touched a neighbor's pact
Then all events would have taken place so rapidly
And so very crazily and haphazardly
That the rich attributes of the world we know
Would not have had the needed time to grow.

Ah, but it is all of the necessitated restraint,
For it ever takes time a scene to paint,
As such as in the unfolding of a leaf,
The endurations for any stepping feat,

As of the emergence of consciousness
And the paused ends of energy's restlessness:
It's of the controlled consequence of collapse
Rather than one that's wholly precipitous.

So now all is known of our heres and nows
Within this parentheses of the eternal boughs,
As well as the why and how of it all has come,
And of our universe's end, but that others become.

Out of energy's dispersion and decay of *quality*
Comes the emergence of growth and complexity.

(The verse lines, being like molecules warmed,
Continually broke apart and reformed
About the rhymes which tried to be nonintrusions,
Eventually all flexibly stabilizing to conclusions.)

THE MYSTERIES OF THE NIGHT

Oh dome of night, spotted with silver stars,
I must ask more than you can grant unto me,
So that thus I might at least obtain that
Which I but wish for in the first place.

I beg you to yield your dearest secrets,
To reveal the full truth of what you are.

Oh, man, I cannot tell thee of all there is,
For I am that, as all that IS—the Wiz,
And as I never began, I earned not my throne,
Yet I reside as the All for reasons unknown.

Much I already know from twilight dreams
And from poems unveiling truth and beauty,
Yet I ask, with my most persuasive looks,
To learn the deepest mysteries of the night.

I have always been, and must be, so jot:
That All is ever here to be, since nothing cannot.

Well then, might lesser answers I obtain, in lieu
Of never us knowing really the why-fore of you?

Oh heavens yes; pose your quandaries,
But ask not immortality, nor youth, nor birth
From my powers of the night, 'though these I have
But know not the why, for I have no First.

Why then, is the universe so extravagant—
With trillions of galaxies of billions of stars,
About which so many planets whirl and twirl,
With so much dust swirling in between worlds?

There are vast multitudes, true, so easily made,
And more; yet they are finite, as must be,
For no cap can be placed on infinity;
If it could, then night would be white with light.

So then, there are stars to burn, as with riches,
But why, really must the largest be so large?

It is because the infinitesimal, the smallest,
Must be so very tiny, so minuscule,

As a simple, continuous function,
Neither composite nor of course complex.

So there is a basic lightness of being
Because anything more would then be of parts
And thus beyond the fundamental arts?

Yes it is that the base can only be as such
When it's just a bit more than nothing;
But there is some more to it; just ask to learn.

Is it too that there are then so many more chances
For arrangements, due to the extravagances?

Not as meant, but that falls out, as it must,
For since the opposite Not cannot be,
I must then be Everything—of possibility.

All at once? Then that is a superposed All.
What makes time begin and then gear its call?

As great as I am, there are two limits
To which even I must ever obey:
My superpositions must either trace back
To total order or to disorder: two.

And so time can only begin from order,
As with matter separated from antimatter—
Time pushed forward by this arrangement,
And further pulled forward by disorder?

'Tis confirmed, with the Big Bang start,
Through the vast stages of diversity,
Unto the end—of entropy's heat death.

As protons to stars to their explosions
And radiations to atoms to cells to life
Unto brains and consciousness?

Yes, from the stars cometh not just our help,
But us too and everything else out there.
All is the continuance of just the one big effect
Of the one big event of the beginning of time.

Earth couldn't be farther out in space, alone;
In all directions it rolls along, unknown.

Epic Thoughts: The Best Of

I look to the stars piercing the depths of time:
They beckon, warm and welcome, the fires of home.

I am that, as the night sky, whom you ask.

I wish that I retain your presence
Within me, in rhythm and resonance.

Everything is part of the IS,
Which is really the best answer to your quiz.

Who am I really talking to?

Yourself, for you are the universe come to life.

I live; I love.

You do not just live; you are life.
You do not just love; you are love.

They are both here.

Life and love do not flee on, just ahead of you, unreachable,
Leaving you but to lean forth and drink their wind.
You are the universe turned around to view itself.

I strive.

Zest, desire, caring, and other feelings sweet
Are your lightning feet for triumphant feats.

I reason.

All manner of shapes haunt the wilderness of the mind,
Many as waste, as in the universe, at large, in kind,
Just waiting and asking to be tamed as sane.

I ponder.

You are the golden chalice to the wine that flows;
Drink, drink!
You are the live and resultant existence that knows.
Think, think!

I imagine.

Thoughts fly in the mind like birds wing the wind;
Imagination is the atmosphere wherein ideas are born
And borne on the waves of the sea in which one sees.

I have arrived, after 13.57 billion years.

All from stardust begins and ends in thee.
The mighty wrecks of the elements are strewn
Across the universe like chaff from the harvest—
Much of the Cosmos a vast wasteland.

Are there others elsewhere as I and all?

Yes, in quite a few places, but afar,
With much intervening space in between.

What more could human mammals want?

This is it.
There is nothing more now, but in future growth.

It is now and I am here.

VACATION PLANETS

Uranus is quite pleasant compared to Pluto.

If you've ever had a dog, you know what I mean;
However, the under-worlded canine has been
Banished from the house of Astro—
To reign as an under-world in the Underworld,
For it's much better to reign in Hell
Than to be an unwelcome guest in the heavens.

Once I was down on Venus,
And the sulfurous emanations
Were so repulsive that any gases from Uranus
Would have been to me as a breath of fresh air.

The gas giant planets' breadth and width is staggering,
And their mooning around is getting out of hand.

That leaves Mars as the only other good place—
Since Klingons have now appeared
On the rings around Uranus.

TO THE DEPTHS OF THE DEEP

Here I stand, holding fast,
Onto my other half.

The zephyr faints, dying in the half-light,
Its caress suspended, as day kisses night,
When for some instants, stretching into moments,
We are neither here nor there but in twilight.

We live at this boundary of day and night,
Our selves merging in the blend of twilight:
You and me, me and you; yours, mine, and ours;
The day-gold melts into the jeweled night.

Above us, fires burn the stars away;
Below us, the Earth turns under our feet;
Within us, unworded dreams haunt the soul;
Around us, night pours blackness on the ground.

Soft and warm, the evening caresses us,
In gentle darkness and quiet stillness.
Here we sense the sweep across the heartstrings,
For we're undistracted by the day's bright noise.

I beg of the night to yield its dearest puzzle,
To reveal the full truth of what it is.

Much I already know, from twilight dreams,
And from poems unveiling truth and beauty
But I ask, with my most inquiring looks,
To know the deepest secrets of the night.

I must ask from the powers of the night
Not immortality nor youth nor birth
But only that I glimpse the enigmatic—
That riddle posed of the conundrum.

...

Follow.

...

The door resisted at first
Then creaked into the crypt,
Powdered rust streaming from the hinges.

Here the answer to All was kept;
But not all was pleasant—it spoke of death,
Of life's end, separate by just a breath.

I saw tombstones overgrown, underswept,
Names unknown—and to all the message saith:

"Read Me",
It said, in words engraved beyond the brink,
"You who live up above: of life go drink;
And you underneath, now lying so dead:
Rest in peace, relax—it's later than you think!"

To learn the Secrets—what IS and ev'r WAS,
One must brave the crypt and ghost of cause.

So into the deep we go, without pause,
To look down, ever down, no self to keep—
Through birth, death, and the shade of sleep,
Through paths unkempt, underswept—

To the deep,
Through the cloudy strife
Of this hazy life,
Through the equations of eternity—
Their non-paternity nor maternity,

Past the realm of the things which seem or are,
Even o'er the steps of the remotest bar.

Down, down,
Where the mind whirls round and round,
As the ear draws forth the sound,
As the eye sees the light,
And of the dark the fright.

Down, down,
Beyond all death, despair, love, and sorrow,
Past yesterday, today, and tomorrow—
The body's guide but the logic of the 'know'.

Down through the fog, the not, and the void,
Where 'God' and everything fail; Oh, zoids!

Down,
Where reigns the night, where the air is thin,

Where the sky and stars are not, but within,
Where the glorious have not their throne,
Where there is one presiding, all alone.

Down, down,
To the fathoms of the cryptic;
Where substance slept with arithmetic,

Toward the spark yet nursed by embers,
To the first and last the universe remembers,
To seek the gem that shines—the wealth of mines,
The jewels so treasured by thee and thine.

What truth accelerates life's momentous gem,
Letting the motto become "Carpe diem"?
Who seized the moment or lost its momentum,
Wearing not the time as its royal diadem?

The World does not pass by—we pass through it live;
Clear your being so the treasure may arrive;
The spirit sparkles of a different light—
The gemstones are of a different mine.

Down, down!
We guide thee, we must carry thee;
We're illumination beside thee...

Down!
Fear not the proof—
It's the beauty of the truth:

Above the ground you were ever born again,
When the roseate hearts were cleansed by dew,
And lucky were you if spring found you new,
As every blossom on the bush blew full.

When these wonders the new morning bestrew,
The beauty of truth was all that you "knew".

Life's hardships there were softened by beauty,
All its weaknesses strengthened by the truth—
As when roses blossomed, like realizations,
Beauty itself bloomed from the well of truth.

For now, rarely enough, existence is left aside,

And yet the essence ever has its other side—
Life, although anguishing, must be lived fully,
Since if you're alive enough to feel its beauty
Then you're exposed to its opposite twin;
Yes, Beauty's other side is Melancholy.

Down, down,
The essence beckons us back home,
As the contained-container is the poem.

When a deep truth is known so intensely
That all of its clothing falls away,
Then one has learned the beauty of truth,
For the reality of meaning is beauty.

When sadness brooded over the morrow,
I once visited the deep well of sorrow.
There enshrined, inseparate, Beauty said,
"'Twas from me that sadness you borrowed."

So do we live the life of art,
Each playing our part?

Nay, that is not life, nor a part, bit,
For there's another dimension to it.
Art and poetry enrich human experience
But they're not substitutes for the living of it.

Like Keat's figures on the urn, blest,
Should we live life any less?

No—because what is deathless is also lifeless!

Down, down!
Truth and beauty must be inseparable,
Although this is seemingly imponderable.

On that sphere above,
Soft breezes ever blew, caressing me and you
As we kissed the roses new and drank their dew.

Reason and passion then merged into one,
As truth and beauty made their rendezvous.

Down, down, ever down—
Through the antiquity, past all of the known—

Arriving at the lowest, remotest throne,
One of the highest perfection,
For it is of the two contrasting directions.

Opposite twins rule the causing call,
The positives and negatives constituting All.

Here the enigma of the ever immortal
Is undone and unloosed through its portal:
The Theory of Everything mortal—
The Idea for which we've opened the door to.

Down, down,
To the end at last!

Here be the lawless and the formless
Of the unordered, uncreated scene.

Here the causeless reigns supreme.

THE WORD—THE LEANINGS
AND THE GLEANINGS

Where in the Woe is Purgatory's bane?
Purgatory's on Venus, where sulfurs rain.

Where in the Heck is that deep Hell of pain?
Hell's found in the sun's heart, oh hot burning pain!

Where in the name of Heaven is Paradisea?
Of Heaven's site no one has any idea—

Really now, where's Heaven one and the same?
It's the world's best kept secret: Earth is its name!

Yes, that's said, but truly, where is the stead
I must tell of them that they're only read;

Of those places spent after we are dead?
It's written of words that language bred.

'Twas hope-word that invented all that was said?
'Twas these that were signed for anything Divine [said].

FLORA SYMBOLICA

A tale I've written, invented, yes, hence
An attempt to unite the Christian pense
With the non-belief, in a middle ground,
Somewhere between mystery and good sense:

With flora mystical and magical,
Eden's botanical garden was blest,
So Eve, taking more than just the Apple,
Plucked off the loveliest of the best.

Thus it's to Eve that we must give our thanks,
For Earth's variety of fruits and plants,
For when she was out of Paradise thrown,
She stole all the flowers we've ever known.

Therewith, through sensuous beauty and grace,
Eve with Adam brought forth the human race,
But our world would never have come to be,
Had not God allowed them His mystery.

When they were banished from His bosom,
Eve saw more than just the Apple Blossom,
And took, on her way through Eden's bowers,
Many wondrous plants and fruitful flowers.

Mighty God, upon seeing this great theft,
At first was angered, but soon smiled and wept,
For human nature was made in His name—
So He had no one but Himself to blame!

Yet still He made ready His thunderbolt,
As His Old Testament wrath cast its vote
To end this experiment gone so wrong—
But then He felt the joy of life's new song.

Eve had all the plants that she could carry;
God in His wisdom grew uncontrary.
Out of Eden she waved the flowered wands,
The seeds spilling upon the barren lands.

God held the lightning bolt already lit,
No longer knowing what to do with it,
So He threw it into the heart of Hell,
Forming of it a place where all was well.

Thus the world from molten fire had birth,
As Hell faded and was turned into Earth.
This He gave to Adam and Eve, with love,
For them and theirs to make a Heaven of.

From His bolt grew the Hawthorn and Bluebell,
And He be damned, for Eve stole these as well!
So He laughed and pretended not to see,
Retreating into eternity.

"So be it," He said, when time was young,
"That such is the life My design has wrung,
For in their souls some part of Me has sprung—
So let them enjoy all the songs I've sung.

"Life was much too easy in Paradise,
And lacked therefore of any real meaning,
For without the lows there can be no highs—
All that remains is a dull flat feeling!

"There's no Devil to blame for their great zest—
This mix of good and bad makes them best!
The human nature that makes them survive,
Also lets them feel very much alive.

"That same beastful soul that makes them glad
Does also make them seem a little bad.
If only I could strip the wrong from right,
But I cannot have the day without the night!"

So it was that with fertile delight Eve
Seeded the lifeless Earth for us to receive.
Though many flowers she had to leave behind,
These we have from the Mother of Mankind:

Eve gathered the amiable Jasmine,
Which soft exhales its breath of friendship,
And by a delicious fragrance in the night
Overpowers the stars with its sweet delight.

The Jasmine impregnates the dew each night
With its friendly perfume of good and right;
Thus morning's incense carries its odour,
Keeping everyone in fresh good humor.

Love's first emotion springs from the Lilac,
For it blooms when Nature is first aroused,
Thus it's love's youngest dream to all come back,
Where it will ne'er again remain unspoused.

When Thyme she sowed, the bees came all abuzz,
And all around it flew their dance of love.
So now we know that those who would savor
The sweets of love mustn't neglect the flower.

Camphire, the scent of Paradise, inspires,
Reminding us to what our soul aspires,
When spontaneous desires overspill,
To tell us of duties we must fulfill.

Daffodils, arranged in their elfin way,
Wear their yellow skirts, like Fairies' Dresses,
And brighten, through the spirit light of morn,
Into the fuller radiance of day.

Butterflies come to life in Pansies' psyches,
Embodied by extension into flight.
They're flowers floating on the air, propelled,
Leaving shadow prints behind on the petals.

The air fills with Honeysuckles' scented nets,
From fairies blowing the honey trumpets,
While they sow vermilion red Geraniums
That grow wild into many countless sums.

The Golden-Throated Lilies sing at morn;
Maiden Flower blushes, its pureness reborn;
Star galaxies of Sunflowers sway,
Echoing the luminosity of day.

She picked some Dandelions ripe enough
To have gone from gold to just so much fluff,
Reminding us, when soft blown with a puff,
That time will spread us too amid the dust.

Chrysanthemums drink the mellow day;
Falling petals carry the light away.
The autumn fog enswirls, the mist upcurls;
Into nothingness the wisp slow unfurls.

Woodbine wets the air with its cooling musk.
Bluebells herald the dim and dewy dusk,
And ring the dance and song of evening knells,
Music tinkling in fairy festivals.

The Evening Primrose only in the night
Opens its cup to drink-in the moonlight,
Then gazes round with silent love and smiles,
Much as we would upon a sleeping child.

Its phosphorescent light guides the flight
Of the flying creatures that love the night.
It looks the swelling moon straight in the sight,
When they make love in the haunt of midnight.

Pearly Everlasting, frozen in time
By Eve's purity, survives cold and rime—
It's a bit of Heaven brought to our clime,
Where it still ignores the knell of Death's chime.

With willowy grace, Eve fished with vines,
And the Willow yet throws out her lines,
As drooping branches that fill the streams
With tears for flowers that we've never seen.

The innocent Daisy, or the "day's eye",
Is a lot like the sun—it cannot die;
It far outlasts every other flower,
Shining even when the sun has no power.

Arbutus too, whose fruits and flowers of
Grew together in inseparable love,
Eve took along with her, as Heaven's boon,
When she felt the kiss of the rising moon.

Out of God's thunderbolt grew the Hawthorn,
On that day when man and Earth were born.
Its snowy blossoms of hope and union
Gave this blesséd world its first communion.

The fleecy Hawthorn sheds its summer snow
To remind us of our birth so long ago.
So Joseph's Hawthorn staff along the way
Still blooms in winter on Christmas Day.

Hawthorn was once known by the name of May,
Its thorns by then having been bred away.
Thus for it the children went a-maying,
And built the maypole, all around it playing.

But the calendar was set back twelve days,
So Mayday was no more! Yet memory stays,
And the Queen of Blossom's day is made
When writers and lovers seek out her shade.

Ever, the immortal Periwinkle,
Which, like the winter stars that twinkle,
Spreads through the snow its glossy flowers,
To remind us of the spring's sunny hours.

Though laughing with all the smiles she wore,
Eve now more serious her burden bore
When she brought forth the mournful Asphodel,
Dedicating it to the souls of Hell.

The Asphodel sustains the Dis dwellers,
Where they rest beyond that fatal river;
There the wretched shades drink forgetfulness,
And to oblivion sink without distress.

Fireweed grows from Hell's sulfurous embers,
As does Purple Loosestrife—dead men's fingers;
But wildflower air revives the dead—so then
Those happy souls can thrive on Earth again.

Quick sprout the Buttercups, all bright and new,
Goblets from which the fairies drink the dew.
From the Eglantine springs poetry's power—
It's the only way to describe this flower!

The Heliotrope turns towards the sun,
Closely tracking its path throughout the day,
But when clouds appear or when day is done
It forgets about the sun and looks away.

Eve brought forth Magnolia's magnificence,
The playful Hyacinth in its sprightly dance,
And Marigolds that follow the summer lost—
Enduring well into the final frost.

From the Poppy we gain full sensation,
Elation, and oblivion's consolation;
When life's miserable pain is too deep
It simulates death with a balmy sleep.

Growing in the cold, near the leafless trees,
Snowdrop bells ring out for friends in need;
They bring hearty hopes to those with hardships—
Icicles changed to flowers by friendships.

Eve carried forth Forget-Me-Not bouquets
That sprouted fast wherever heroes fell;
They bring back all of the happiest days
To sound in our hearts as memory's bell.

Holly, the harbinger of spring desires,
Blooms all winter long, and with hope inspires
Our cold and dreary hearts to chime and ring
With good cheer and love for everything.

She took poisonous Foxglove and Nightshade
To balance with woe the good that she gave,
Offset by Amaranth, which if kept in shade,
Would not even after death ever fade.

And for the romantic art, Cupid's Dart,
To spur men and women to make their move.
Connected by Nature's arrow of love,
They deep impart the passion of the heart.

And Coral Bells, rung by bees and hum-birds—
A melody of tones without the words,
And airy sprays of frothy Baby's Breath—
Gurgling with all that's much too sweet to purge.

There, sweet spikes of aromatic Lavender—
Ready potpourri from Heaven's splendor,
As all around lay the symbolic flowers—
To soft drowse the spirits into slumber.

Yet more we know from myth, lore, and legend,
Of flowers that gemmed the fields of Eden,
And from symbols and wisdom handed down
Through oral tradition in floral towns.

Wherever Eve breathed sprung floral dreams;
Ever she walked water followed in streams;
'Ere she wept, tears bedewed the Earth in bloom—
A Cedar tree even grew from her tomb.

'Dead' flowers are reborn by Spring's breath:
An ethereal floral wonderland
Of everlasting recollections, and
Some even retain their color after death,

Like Amaranth, as mentioned earlier,
Or Lasting Beauty, whose secret elixir
Grants us flowers red through a year of days—
Oh but that life and love would never fade!

Or Cedar, "life from the dead", the emblem
Of eternity, as the preservation
Used for mummy cases and carved figures
That last 'forever': immortal rigor.

Tracking Eve's trail throughout the ageless years,
We find Lady's Slippers, Lady's Fingers,
And Lady's Smock—all parts of Madonna,
Her whole self, in fact, in Belladonna.

She wore a chaplet of sweetening buds
That burst in bloom when fed by air and mud,
And a garland of sprouts to strew about,
With a rosary of shoots to put out.

She scattered the Fern's seed at midnight's peal,
To ask that treasures of the Earth would reveal
The flowers of woods, waysides, and shorelines—
All remembered by florigraphic signs.

Eve planted the Tree of Life, from which we
Could obtain lumber, fuel, and homes, for free,
Plus weapons, wood, tools, food, and medicine—
To mold the Earth into a place we could live in.

And Clover bushes, the haunt of the bee,
Bamboo grass too, for home and social need,
And Lumeria, whose transparent seed
Looks much like the moon, in all honesty.

Continual Morning-Glories each dawn
Guarantee that day will always come on.
Bindweed and Honeysuckle yet entwist,
To tell us that lovers will ever persist.

The melancholy Thistle is a cure
For the blues when taken with wine that's pure.
Chicory in blossoms maroon is clad,
Its young and tender leaves used for salad.

Eve gave freshness, fragrance, to the Lily,
And seized Hemlock, the Devil's property,
Left us Hawkweed to clear the sight and wits,
And brought Hellebore to purge evil spirits.

The Hawthorn, here yet again, blooms redux,
Like Blackthorn in Christ's crown, as thorns do,
Or as wood of the true cross where He died—
All seem to miraculously multiply!

Eve's saplings drank of the Earth's gushing breast,
And produced the primeval forest.
Somewhere this secret wood remains, unguessed,
The place where all man's sorrows come to rest.

Life is a flower whose leaf is summer green,
Whose spring was purple passion Eglantine.
Although fall's second spring may intervene,
The frost at last is the winter seen.

All Earthly pleasures dear to us Eve brought,
Provided by the Master's afterthought:
Honey, juices, syrups—all hand wrought,
Nuts, berries, and fruits—nothing went for naught.

Eden's sinful Apple, the cause of it,
Made for harsh apple cider, but when it
Was heated with sulfurous brimstone it
Then turned smooth, the Hell taken out of it!

The Clematis, with its clinging habit,
Makes shade of Travelers Joy at inn porches
For wayfarers wearied, warm, or unfit;
Its leaves are the clouds, its fruit: star torches.

From Quinine, medicine that could relieve;
Of Citron, cure for snakebite—death's reprieve;
The Ginseng refreshes memory's streams,
Calms the passions, and begets pleasant dreams.

Basil Leaf is a ticket to rapture,
Passion Flower, to atonement—a day-star,
And Yew, the oldest living thing on Earth,
Yet remains alive—six thousand years worth.

The Trefoil, for love, heroism, and wit,
Grants power o'er the banshees of moor and pit,
Who would steal the soul, and against all snakes
Poisonous—they scuttle into the lakes!

Edelweiss, a white flower most gallant,
Is the heart left by an angel visitant.
Mistletoe lends a green indoor refuge
To the wintering spirits of the wood.

The dusk deepens, night's pot of tea steepens;
Silence descends, as when a gift opens;
Eventide rises. On high, Orion camps.
The eyes catch stars like fireflies in lamps.

Our shadows are touching, in the same shade—
We embody, in third dimension made;
We kiss, drift, cross into each other's role;
Spirits open—rainbows meld in the soul.

If Nightshade you eat you'll become as so,
And will see the ghosts, shades, and dark shadows
Of those who came before our humankind,
Those whose spirit-worlds overlap the mind.

The Tuberose is a dangerous pleasure,
Even when taken in but small measure:
Its exquisite scent has such great power
That it can wither you within the hour.

What's that? Phantoms that are but a glimmer
Of the life and light of some halfway scene.
Of beings twixt man and angel, they shimmer,
As one might remember them from a dream.

They, cupid like, are the souls of flowers,
And wear petal cloaks and have wings that blur.
They sleep in Cowslips, where with childhood's ear,
You, listening, all their music can hear.

They're sylphs, tree spirits, wood folk, and fays
Gathered in posies of living bouquets.
Knowing well the language of the flowers,
They bestow their favors on the growers.

There's a tunnel back to Eden's Garden,
A funnel, really—our small end open,
And through this fairyland we'll return, free,
To hang Adam's Apple back on the tree.

Sprites shadowed Adam's Eve throughout the land,
The seeds sprouting everywhere by their hand,
The growth blessed by a pixie's twinkling wand
That showered the plants with a fine dewy sand.

The naiads too spread germinating seeds,
Among them these many blossoming deeds:
Perpetual-Flowering Carnations,
And sparkling Buttercup potions, as in

The silken saucers for Hollyhock tea,
In which a child could capture the wild bee,
To hear the aggravated buzz, in play,
Then unstung, free the bee to fly away.

The Elves grew Basil, Wolf's-Bane, Cucumber,
Cinquefoil, Meadow-Saffron, and Germander,
Even Gillyflower and Primroses,
To which the fays gave their dewy kisses.

Cotton grew, woven by the wee people
Into clothes, with a whirling spinning wheel,
Whose spindle was the stinger of a bee,
Weavings that surpassed the spider's best web.

Fireflies followed and lit the way for the
Little weavers who were chased by jealous
Spiders; the folk hid in a Cotton ball,
The spider finding nothing there at all.

The weed flowers came, marking autumn's track,
The blossoms that almost brought the spring back,
But winter's white death wrap was drawn over,
Smothering the earth's last warm sweet odour.

Such then comes the end of summer's dreams,
The blanching of the grassy banks of streams,
But all fragrances the elves remember
Through their sleep during the winter embers.

Youth and Beauty made agèd Winter mourn
For Summer's grain—the waving wheat and corn,
For Old Autumn, withered, wan, had passed on,
Leaving the earth a widow, weather worn.

The blossoms fall, showers of fragrant beauty,
As leaves fade while the bulbs store up energy;
Faeries' floral dreams grant this destiny,
For these leavings enrich earth's potpourri.

Flowers lay their heads to sleep in soft beds,
Blanketed by webs of gossamer threads;
The fairy creatures cast their spectral glow,
As winter stars—floral twins—start to grow.

Later, when surely all the world is dead,
A fairy stands atop Old Winter's grave
And says, "'tis not dead", and by magic bred
Makes Snowdrops flower in the tomb's heat wave.

Winter Aconite, an early flower,
Grows even under the season's dim power,
And its bright corollas far out-splendor
The winter sun's pale and paltry color.

Nymphs slide from their cocoons, their pinions
Yet wrapped and wet, then breathe the earthy air
That calls them forth into life's dominion
To fly and flutter in flux here and there.

Flowers spring from the footfalls of a lass;
Foliage withers where evil spirits pass;
But where unknown colors shine fairies mass,
And drink the twilight dew off of the grass.

The elves blow their pipes to awaken
Nature's Flora, that her step may quicken—
And from the odours memories recur,
As we're given back our youth of summer.

The blooms are a crimson mist, in green blade,
Through yellow air, beyond the deep blue shade.
A white mist drifts through azure skies, bade
Toward purple mountains—fragrance of the glade.

In the spirit world, the grass is greener,
The hearts redder, and the passions pinker—
Orange, Cherry, and Violet are planted colors,
And twixt blue and green falls a new tincture.

Petunias grow wherever rainbows touch,
Their colors vibrant, a bouquet as such
Of rays that make the flowers glow so much:
Heaven's prismatic radiance—life's clutch.

Love is reason enough for its giving,
For beauty is its own excuse for being.
The doing of good becomes its own reward,
For the truth does best define its meaning.

In the luminous backwood haunts night plants
Are seen growing fast from the touch of nymphs:
Fairy's Frocks, made of elfin sowing—of
Heart-halves of Lady's Lockets joined in love.

At night, Tulip lamps light the lover's gate,
As Hollyhock torches illuminate.
The secret hollows glow from Crocuses—
They're cups of sunlight stored for the muses.

At woodland's edge, wee folk leave sentinels,
The Bugle flowers, to announce to dells
The entrance of lovers into the wood,
So all can enjoy the amorous mood.

Wherever the elves themselves have romance,
Wild Pansies, known as Jump-Up-and-Kiss-Me,
Spring from the power of their loving dance—
Emanations from the sprites' imagery.

The eyes love to rest on the sky of blue,
While Eve upon the greensward smiles at you—
A new life colors the world in between
Devils and Angels: Earth's human pristine.

Eve set tufts of Anemones, fully blown,
Ever after given as the wind's own,
And vines, wreathing and twining, overgrown,
And odoriferous blooms in bunches sown.

Across the lea and on the moor she shows.
Along the lane and through woodland meadows,
Eve—Mother Nature—yet lives in boughs
And thickets, still imparting all she knows.

Some flowers close, protecting their pollen
By "sleeping", some at morn, some at even,
Some at other flower-clock hours—*somewhen;*
And some, like Jewelweed, never open.

The glow worms, fairy stars come down to ground,
Gleam the shadowy woods through summer's round;
Then fall's leaves flutter through the quiet air,
The autumn being the sunset of the year.

Brown is Death's coloring of all that grows,
So faeries don't allow it in their rainbows,
But beyond the spectrum, where we can't see,
New hues paint their phantom activity.

Elves find Venus shining in broad daylight,
Knowing where to look as if it were night,
Then follow her as the evening star,
Till with her fiery lover she takes flight.

Just before dawn, amid the dew and moss,
Elves ride on a moonbeam made of Bugloss,
And see the North Star and the Southern Cross
In the same sky, 'most all the way across.

Now the Earth is very old, but each spring it
Turns young again when nature reinvents it,
Constructing the Temple of Flora outside,
In desert, field, wetland, woodland, and wayside.

Spring kisses the earth, leaving flowers there,
Like those whose perfume first scented virgin air,
As again, the fragrant glen, in Heaven's prayer,
Hails Earth's anniversary with flowers fair.

Slake love's thirst for life's earthly endeavor
Near a stream where wildflowers grow forever.
Flowers influence our feelings—deep they roam:
Flora's fairest flowers compose Heaven's poem.

The pure white flowers of Paradisea grow
Only within the sub-alpine meadow,
Not to mention Sundrop, Saffron, Twinflower,
Pomander, and a thousand other flowers.

For supper, Eve savored salad made from
Thyme, Mallow, Bibleleaf, and Sugarplum,
All edible and flavorful flowers,
Mixed with Chervil, Lovage, and Sunflower.

The Lavender, Rosemary, and Sage all
Release fragrance when crushed by a footfall,
So herbs are strewn on floors to clean and scent:
Odoured ornaments preventing aliments.

Early Sage, before it became dilute,
Kept man immortal—an ever-green root.
Though now diminished in its once great power
It still keeps us healthful in summer's bower.

The Crown Imperial refused to hang
Its head at the foot of the cross, so vain
And proud of its majestic reign—so now
Its petals must droop and weep nectar rain.

Heaven's patron of arts, grace, and license
Left us sweet-smelling plants with flowered scents
And aromas redolent—florescence
In flush and prime of days reminiscent.

Blooms have eternal life in Heaven's glade—
An ethereal floral wonderland
Of everlasting recollections;
Oh but that mortal life would never fade!

When Eden fell, all elfin creatures too
Were loosed with Eve into the world anew.
They're tenders of the precious flowers few,
Of the flora that in the Garden grew.

There! What uncanny things flock, in between,
Unknown in the shadows, there but unseen?
They're dream-visions—completing the triad of
Earth's Heavenly things, with flowers and love.

Breathe flowered air and you'll never know death,
Your incarnate life an eternal wreath.
Breathe ambrosial incense, balm, and spice
Of flowers as fragrant as a fairy's breath.

Eve's elves gave us the taste of Strawberry,
The messages of the Honeysuckle,
The signals of Wisteria, and the once
Neglected memories of Rosemary,

And the sweet breath of purple Violets
With the enamored voice of rivulets,
And Scarlet Pimpernels, that aft nice days pass,
Enfold—they are the poor man's weather-glass!

And brilliant clumps of Blue Delphiniums,
Soft Irises and sharp Nasturtiums,
Dewy-eyed Pensings, velvet smooth and dear,
And Lilies of the Valley—they're Eve's tears.

Eve carried Myrtle too, meaning perfume,
To rouse Beauty from her watery tomb:
Myrtilla rose from the sea in old Greece,
Adding Myrtle sprigs to the laurel wreath.

The arts were first born from the Acanthus,
In the wreaths of it made at tournaments—
They're engraved in the columns of Corinth
As Greek architectural ornaments.

Vervain too, with the power that enchants;
It brings on visions of a sweet romance,
Gathered as Druids did, by inner sight,
When Sirius rose against the moonless night.

Orange Blossoms are generosity's shower,
Being at once fruit, foliage, and flower.
They bear the legendary apples golden—
Often guarded by a ne'er-sleeping dragon.

For remembrance, Eve brought us Rosemary,
The Lily too, white for its purity,
And the Tulip, which does declare its love
By the truth which it is the beauty of.

But all the flowers mentioned herein above
Would not have made this life worthy of,
So Eve took the Rose—the bloom of love,
Right under the eyes of Heaven above.

The Rose was pure white when it first was born,
Until she kissed it with her ruby lips—
Or 'came it red when Venus fell on a thorn,
Rushing to the aid of struck Adonis?

Or did the Rose sprout forth, all fully blown,
From the heart of a Goddess, do you think?
Or was it out of Cupid's nectar grown,
When he poured to Earth that Heavenly drink?

Or when the nightingale, with hope forlorn,
Overpowered by the Rose's perfume,
Impaled itself in love upon her thorn,
Then revived in the beauty of the bloom?

With the Rose the Earth is rich forever;
It's born from spring's dying kiss to summer,
And wears all the gems that the dew has wreathed,
Blooming wherever summer's breath has breathed.

The winds make love to the flowers of May—
The woods burst with the joy of Eve's bouquet!
Like Flora we too from Eden have come;
From all that's gone before we are the sum.

Now Heaven's favors are spread all around,
For the flowers, fully blossomed and grown,
Wave and smile, as miracles from the ground—
Reminding us all of what love has sown.

FREE/FIXED WILL

Ah, in the whole, you're just afraid of being unfree,
But, hey, look, behold! There is still so much beauty!
It's a sublime law, indeed,
Otherwise what beauty could there be?
So here the coin's other side speaks—
A toss up, weighted equally.

It's from the searched finding of truth—not of fright,
Though determinism is really not a very pretty sight.
Beauty exists either way, for there is still novelty,
But 'determined's opposite is of an impossible currency.

How dare you curse the freedom to be;
It's because you are scared of He!
What greater proof of inner freedom then
Could His gift of wild flight to us send?

Really, it not of a scare that He is there,
But because 'random' cannot even be there,
For, then on nothing would things depend—all bare,
If it could even be, but it has no clothes to wear.

I swear I am more—that I do act freely!
Don't pass off my passions so calculatingly.
I'll let the rams butt their heads together;
One absolute position subsides for its brother!

Yes, it seems that we can choose, even otherwise,
But what is within, as the state of being wise,
Knows not the hidden, non-apparent states below,
For that is a 'second story', having only one window.

One rigid mode of thought' score
Consumes the other with folklore,
Unbending, unyielding with perfect defense,
To orchestrate life's symphony at the song's expense.

We're happy to just find out the truth;
However, when subjected to the proof,
We wish that the coin could stand on its edge,
And see that it cannot, which is knowledge.

So lets define the world and human existence
On a couple hundred years of material witness,

Or burn the measuring eye to the stake!
After all, our freedom's what it seeks to forsake!

Evolution didn't work by chance for us to live,
For natural selection is the scientific alternative
To Intelligent Design from something outside;
The coin of determination has no other side.

The secret is simply that a secret does exist
And no amount of data can take away this,
But this doesn't mean a ghost in the machinery;
But perhaps the heart isn't just a pump, the liver a refinery.

We often forget the secret, willingly,
In order to live life excitingly,
Which it still would be, either way,
As we're still part of the play, anyway.

But of course there is a past of 'whethers',
Through which we've been weathered.
Surely we are moved as dust from gust to gust,
But is two-twice-two as four always a must?

Math, too, is a must, and we try, as ever,
To predict a week ahead the weather,
Yet the data seem to much to work with,
But indetermination measures not random's width.

Is not an unfree will a blatant contradiction
Developed from the an 'enlightened conviction'?
If I've made a choice then I have willed it
And if it's been willed then freedom's fulfilled it.

This what I mean, that the will willed one's self,
Which is that one does not will the will itself.
The neurons vote, based on who one is—
No one else is there to answer the quiz.

And of course it's in and of a misguided pit
To say that from the past we've distilled it.
Is not the idea of complete self-autonomy a ruse
Born from the illusion of the existentialist blues?

We distill what comes into us, too,
For it has to become part of us, new,
For mirror neurons act it out, while we are still,

Invading our sanctum and altering the will.

But of course, this is to be much expected
From a culture that lacks all mythical perspective.
'Nonsense' we call it, a virtue of not thinking,
From which we have long since been departing,
So now will behold in all its transparency
Beyond childish ideals of essence and archaic fantasy.

That's close, but it's thinking that has grown,
By science and logic informed from reason sown,
In place of feeling, sensation, wishes, and the pleas
To have the universe be what it ought to be.

Do not distort with a desire for meaning.
Oh, the babe, lets leave the child a'weening,
But I ask of you: have you not tried in-betweening?

There are two ways of living, sometimes merging,
One of just 'state of being', of its only showing,
And one of the being plus the under-knowing;
As with our life's wife, we dwell not on hormoning.

And in that same breath we say all is forgiven;
Why hold humans responsible, leading to derision?
Of course an eye for an eye was an unjust decision.

Well, we have a system that draws a line between
A crime of passion and a thought-out, sought-for infliction.

"The universe made me do it," says the accused,
And the Judge replies, "Well, this does excuse,
But I still have to sentence you to the pen,
Until the universe can't make you do it again."

Why must it be a question of absolute freedom
As complete randomness over an unbending system
That structures everything that ever was, is, and will be,
Right down to the elementary structures
Of incomprehensibility.

What is set forth in the beginning
Is ever of itself continuing,
Restrained by time, yes, but unfolding,
For there is nothing else inputting.

I may understand why this has to be;
I have felt the rapture of black and white toxicity,
But why subjugate all possibility for novelty?

It will still be novel, even such as a new parking lot,
For the dopamine neurotransmitters will stir the pot.
New is still new, on the grand tour through life;
Then do some predicting, to then avoid some strife.

Can such a thought hope to cast a wrench into these gears,
A tool so heavy that dissuades all of our fears?
Will all order and inertia be torn asunder?
Will we have giant ants wearing top hats over,
With all rationality considered a blunder?

The truth was not sought to drop a spanner into the works,
But it even turns out to grant more of compassion's perks
For those afflicted with the inability for learning,
Thus eliminating the great annoyances burning.

Am I simply a delusional puddle here,
Perceiving just my liquid perimeter,
As I think to myself I can control
The very rain that expands my rule.
And the humidity that thins
Should I condemn as that which sins?

There are no sins, but just destiny's fate,
Which even includes one's learnings of late.
We and all are but whirl-pools, of the same oscillations,
Some lasting longer, yes, but of the same instantiations.

Outputs without inputs cannot ever be,
Or the actions would pop randomly,
Yet to some people that's the enemy,
A useless state that's not here, thankfully.

AUSTINO'S HOLY QUEST

The golden dome of Sancta Sophia shone,
Along Saint Austin's journey from olden Rome,
It reflecting the flaxen sun's blonding band,
He unto Britain to spread the Word to England.

Saint Gregory, who was to have gone this far from home
Had been elected Pope and so was recalled to the throne.
Saint Austin was ever a lively and wandering monk:
A French-Italian nun packed forth into his trunk.

A carrier pigeon flew the latest to Austin's leg,
From his friend the Pope, known to him as Greg:
We shall conquer not with our Papal Army's spear,
But with the requisite Christianity, dear within its fear.

Saint Austin was taking his time, seeing all the sights,
With the nun ever out of her habit during the nights.
It was like Google's stunning street level view,
As observing everything first hand as always new.

The world turns, shadows fly, energy spreads apace,
Consolidating here and there, not an impatient race.
Muslims crash the gates, and now new Rome is cleft,
A prism of many-splendored light is all that's left.

Austin mends his detour and through the forest flies,
Recalling Galileo's last look unto the starry skies.
Sister Angelina lifts her cup, the alpha and the omega,
As they raise their glowing minds to the mirror of Vega.

Once the great Roman Empire was everywhere known—
The center of Earth the center of the solar system's home,
While the saints walked along, bearing the world alone,
Yet life and play must precede the spreading of the tome.

They would be gone for over twenty years, but not forty,
For they would ask directions, unlike Moses and his party.
God had sent a plague to Europe, it killing one in three,
But none for a year now, yet they dallied one more to see.

The Alps were climbed, soaring into the realm of Heaven,
And there they stuck their heads into that glorious haven,
But soon ducked them back down again, just in time,
For there was a big soccer game going on in that clime.

Did they have to bring the invisibility disorder unto Britain,
Until all the rampaging residents there were by it smitten?
Why should claims of unknown essence precede existence?
What good would this hardy toil bring to the persistence?

In France they lay among the harvested grape crops,
Loving and squeezing out the best and last drops.
All was so lush and green, this world as seen,
And so they lived of life and love in-between.

Pope Gregory was a scientist through and through,
Sending them to spread the word of God all anew—
Gregory's continuing cover over gathering resources
To manage all of his realistic experimental courses.

Many years later they took a boat across the channel,
Arriving in the mixed up country of heathen rebels,
Then thought: if weakness can be turned to strengthness,
Then we have to tell ourselves that's another weakness.

Oh there was ever the word of the ancients to pass on,
Although tending to scoff at the beliefs of the ancients yon.
But one can't scoff at them personally, you do see,
To their faces, and this is what annoys you and me.

However, the first thing learned was to forgive one's-self.
Then told one's self, that self of the more learned self,
Go ahead and do whatever you want, it's okay to be,
For there is this lovely parentheses here within eternity.

Life is a constant battle between the heart and the brain,
Fought amid the bright sun, the darkness, and the rain,
Pondering this and that, what be, on and on anon,
But guess who always wins? It's ever the skeleton.

Yet this is the story of St. Austin who wholly brought fore
Christendom to England, but they don't like it anymore.
Religion will eventually fade, yes, but in the interim
It will change, an initiation of its slide into oblivion.

The stodgy elevation of doctrine over ethics
Will no longer carry the day, and there will be less
Emphasis on believing, with more on belonging.
All will become more democratic, with much singing.

So then when Saint Austin came to his staff
And pulled it out of the earth, onto the path,
Incontinent by the might of our Lord's mountain, [what!]
Sourded and sprang there soon a fair well or fountain.

He and all were bathed of clear and flowing water's sip,
Which refreshed him well and all his fellowship.
His staff was cured and he ever guarded it more closely,
And so the years went by, alive and well with being, mostly.

St. Austin returned to Rome and its discussion forum
Twenty years later, as 40 would have been too long for him.
Pope Gregory welcomed him, *"Has it been so very long?"*
"Yes, for I stopped to smell the roses the way along."

*"Good. So, we have conquered Britain wide and long
Through our Christianity's deep and probing prong,
Rather than with Papal armies that are long since gone?"*
"Yes, it has been done, at least until Darwin comes along."

*"I see that you long ago posted about the flagellum,
Eliminating Behe's claim to irreducible complexity."*
"Yes it's very old hat now, but one can find it via Google."

*"And good old Behe didn't come through, so high,
At the Dover trials either. Well, so much for that guy!"*
"Yes, and now the believers don't want to believe it—
What I said about existence just because I said it."

*"The idea is that they only address the ideas presented,
For proof or refutal—no worry about who it represented."*
"Yes, so what's new in this territory?"
"I'm now finishing my inflation theory."

"Inflation was so rapid that the particles in pairs
Of the always temporarily emitted virtuals there
Were forced to become separated from one another,
With some then to remain as enduring rather."

*"And so they made a universe!
That does it then; it's the last verse.
I'm declaring it infallible, a snap,
Even before the next WMAP!"*

"The Bible will be seen to be of but human construction,
A result of human instinct, frailty, fear, and no wisdom;

So people actively speaking to each other, with laughter,
Will come to replace the passive readings from scripture."

"May the quest be with you too, the life and the fun."
"Let's go have some beers and smokes with the nuns."
"Tobacco hasn't been invented yet, oh darn it."
"Ok, let's go down to the lab and work on it."

THE RESTLESS WIND

Rising slowly from the cold dark hollows
Where the night airs fell and soundly slept
The restless wind left her secret bower,
And gaining strength, lovingly surrounded

And caressed the willow trees, which wavered
And swooned in her wake, as she, the wild and
Wandering wind, flew by in a cool breeze
From the west on her undulating wings,

Spreading the incense of the morning to
Nature's world of growing and living things.
She woke the flowers from their slumber
By drinking from them their blanket of dew,

Then told the tales of the joyous forest
To the birds, who soon carried them aloft,
Thence into my ears: songs of streams flowing
Freely and stories of a glowing sky,

That promised many sunny hours to come
In the dreams of those who felt her passing,
As sleep was washed from their languid eyes
When they sensed that new dawn arriving,

As if some transparent veil had lifted,
When she gently stirred the embers of the
Last watch-fire and whispered softly to them
That the stars had gone and day had begun.

NOSTALGIC NOTIONS

I ran my hand along a picket fence,
Counting heartbeats and running like a child,
Still carefully not stepping on the cracks,
Noting the furrowed ants bustling, thriving.

I wondered at a old chestnut tree that
Had somehow survived the blight, it towering
And ever so gently tilting the walking plane—
Presenting me with more ancient notions:

Of tire swings, swaying, hung from low branches,
And of lemonade stands, secure in the shade.
My youth came flooding back to me, into me,
And so I continued to give it life:

The back door of a bread wagon opened,
Releasing the fresh-baked aroma;
Mother came out with a handful of dimes,
Buying what would've taken three hours to bake.

On the houses' steps rested newspapers
And the sturdy rounded bottles of milk,
Compliments of Elsie the cow, truly
A vision from the grazings of childhood.

We played games on these walkways, like hopscotch,
Roller skating, and marbles. My bag of jewels:
A cool green cat's-eye, a big blue boulder,
And varicolored pockmarked throwaways.

There—a lush garden lovingly attended
By an old lady, accompanied by bees
And butterflies, all of which caused further
Indulgences in my flights of fancy:

As children—and now if we're young at heart,
We'd pause in play when that first butterfly
Fluttered by, that fragile ephemeral
Vision of something almost heavenly—

A flower floating on the air, perhaps,
Signaling that our endless summer had
Begun, that something called "school" was now an
Artifact of ancient history.

Did the butterfly first arise from the
Soul of the pansy, before human times—
One of those edenesque transformations
That is inexplicably magical?

The metamorphosis is still charming,
Albeit but from a caterpillar;
Amazingly, delicate as they seem,
They flutter all the way to Mexico,

And take their sweet time, alighting here-there,
Meandering from plant to bush to flower:
We learn that there's more fun along the way—
The journey as rewarding as "getting there".

During our carefree days we'd swim the pool,
Diving off the side—after pennies thrown,
Retrieving them from the bottomless deep
Near the big drains—then rising up, breathless.

Still at the garden, my mind back from flight,
The gardener beckoned me inward, and
I leaned over the fence to smell a flower,
And a thousand memories reoccurred:

Each Morning Glory blossom lives but for
A single day, and is replaced by
Another, each in succession shining
In its morning glory, wilting in noon's heat—

Withering quickly in the afternoon,
Then languishing throughout the evening—
Their happy message to us being that
Another day will always come on.

The Amaranth intrigues—its leaves never fade,
Even long after death, ever remaining
Vivid red; could it be, somehow, that a
Portion of the immortal lives on there?

There, the blinding luminosity of
Sunflowers; we dried the seeds and ate them,
Each still a glowing ember of memory
Of the bright days among a thousand suns.

I drank up Buttercup portions of the
Bright yellow light from the elfin goblets—
And entered the realm of fairies, pixies,
Fays, trolls, goblins, brownies, gremlins, and sprites.

We had cherries, and a grape arbor, too—
Eating them fresh, competing with squirrels
And birds, always forgetting to wash them,
So sour they were, then spit out the seeds.

I walked on, and saw a lake surrounded by
Old and broken down vacation cabins.
Of course we were never "there yet" when we
Asked, but soon dozed off, tired of asking.

We dug the worms at night, keeping them moist,
And got up with the sun to fish, and then
Skinned them and cooked them for lunch or dinner—
This to me is America Remembered.

Dad was always out fishing—my brother too,
And me less often. Now I clearly see
That fishing has little to do with fish,
But with cool breezes, moist air, peace, and quiet.

I wore my life preserver all day long;
Once I leaned over for a closer look
And fell in, swimming with the fishes,
Then pushed up, my life jacket now broken in.

We puttered to a mysterious island;
And there we found—nothing, but camped for lunch,
Feeling like pirates, and telling no one
About it until a whole day later.

At night we watched the bears forage for scraps
At the garbage dump; however, one night
The bins were empty when the bears came out—
Then they all turned and looked over at us!

Mom used to say "Come in out of the rain",
But nowadays, the sun is dangerous,
Unless we wear sunblock, so she says,
"Have enough sense to get out of the sun!"

After a storm, when the sun returned,
We'd run out to see if there was a rainbow—
That shimmering otherworldly vision of
The colorful secret of simple white light.

How are colors made from three primaries?
Why is the sky blue? What unknown colors hide?
Well, color was invented in the 60's;
Just look at TV shows made before then!

To keep cool we once carried pinwheels, fans,
Parasols, and sucked on a piece of ice.
Now with TV, internet, and cooled air,
We stay inside of the house all day long.

By eavesdropping on the party line, we could
Hear real scandals and idle rumors, and,
If it was more interesting than watching
The grass grow, we stayed to hear the whole story.

Before the invention of the telephone,
All was conveyed by tell-a-woman, but now
We only answer to computers, saying,
"To talk to a human being please hang up".

The corner market carried everything;
Eden's shiny red apples called out,
"Touch me, take me, eat me", and soon trouble
Was at hand but it was crispy, sweet.

I rode my bike everywhere; I always crashed
On the killer hill, on roller skates too;
Now I drive my car there, carefully;
Yes I'm finally getting over the hill!

Always picked up a penny for good luck,
And pins too, for even more good fortune.
I found a horseshoe all of the sudden—
'Twas bad luck 'twas still on the horse's foot!

Rural cemeteries were as parks back then,
So we played near the duck pond, giving them bread.
Some years later I returned, like a goose
That had been away for too many summers.

There were monkey bars for the climbers and
Seesaws tottering for the restless, and
A refreshing sprinkler to cool off in, but
There was always some kid sitting on it.

We made greeting cards, keepsakes, with ribbons,
Lace, assorted scraps, and original words.
Now we buy ready made cards with fluffy words—
In a day or two they are in the trash.

Simple pleasures are as free as ever:
The sights, sounds, and scents of nature; picnics,
Reading, writing, giving, riding, playing—
It's hard to ever get bored, isn't it!

THE YEAR

Hail!
Winter storms the Year
In the month of Brand-new-airy,
Then Feb-buries us in snow!

March, Lady April! Spring!—
Let's reign as we May
With sum(mer)maids
Named June and Ju-lie,

Until, after A-gust of
Hot withering wind,
The sunny fire burns out—
'Cept embers, when
Leaves Fall into Oct-tomb-burr—

Till—no leaves, No sunlight,
No sky, no warmth—
No-vember!

Next de rain, de sleet,
De cold—De-cember,
When all that we can do
Is but sweet Remember.

COLOR SYMBOLS

In the nether world, I learned the lore and
Legends of the colors, of their uses
In nature and emotions, the whatfor
Of their light's glowing activity:

All color variants, quite numberless,
Are made from the three primaries, no less;
Namely: red, yellow, and blue—often backed
By colorless white tinges or shades of black.

From just these three essential hues derives
All the heaven's prismatic radiance,
Myriad colors of floral brilliance,
And technicolors that seem so alive.

The offspring of married red and yellow
Is the secondary, orange, a bright fellow;
Its sibling, of blue and yellow, is green,
With of course some gradation in between.

Saintly brother purple, twixt reds and blues,
Completes the second generation hues.
Next to arrive, lime-green, is a grandchild,
As are all the tertiary colors wild—

They're crimson, magenta, maroon, scarlet,
Amber, auburn, salmon, ocher, russet,
Mauve, taupe, fuchsia, cherry, cerise, umber,
Teal, emerald, and vermilion others.

Strangely enough, all the color-pairs
That symbolize seasons and festive fairs,
As they're found naturally in nature's ways,
Do contrast on the color wheel, crossways:

Direct opposites on the color wheel,
Sky-blue and leafy-orange represent fall,
For they are autumn's contrasting colors—
That quite up for its lack of flowers.

As with crocus, spring's floral colors yet
Remain yellow primrose, purple violet—
The sensual sun, as it were, warming
The virginal earth, with love, into spring.

The Christmas Holiday Season's scene
Is of opposing hues of red and green—
As in Holly, berry-red, ever-green,
Or in Poinsettias' red flush, leaf of green.

We're out of diametric color sets,
So which for summer? It must then contain
The entire spectrum, as these the sunset
And the rainbow express in shine and rain.

Since winter's snow hides all things out of sight,
Its colors are hidden inside white and night—
The cold season's symbols, for they conceal
All of spring's and summer's bright floral feel.

For that as different as day and night,
We have the twin-opposites: black and white;
For the day-clock first became dark and light
When twin-gods split day and night, wrong and right.

Heaven's splendor, white, for purity, bless,
Holds all the colors of prismatic light,
But the symbol of the Prince of Darkness,
Black, removes all the colors from our sight.

So then, it is proved that in both nature
And in the color wheel opposites attract
And complement in their contrast—to procure
Both real and symbolic color contracts.

Next we'll turn to the colors lone, to see
The whatfor of their light's activity,
But first, let's ask, *Are there any missing hues,*
Unknown, hidden in rainbows, or not used?

Hidden colors? No, for I see how red goes
To orange, graduating through the rainbow
Into yellow and on through green, to let
Blue into indigo to become violet.

Perhaps between green and blue lies some new
Tincture unique enough to be it's own hue,
But alas, those turquoise waves everyday
In tropic seas wash that theory away.

Yet there may be some new colors that lie
Before or beyond the spectrum and the eye,
Like infrared or ultraviolet,
Or gold, which only the fairies can see.

But what of clear, white, silver, gray, or black?
Well, they're not true colors, for, either they lack
All color (black, clear) or hide all hues (white)
Or are mixtures (gray, silver): black-white.

But wait, there is a well-known color,
One quite common in both dress and nature,
That cannot be found in the rainbow—
Give up? It's brown—and has nowhere to go!

Brown is the color of death, like the leaves
That crumble dry and lifeless when earth grieves,
Which is why the faeries won't let it show
In their magically spectral rainbow.

But alas, brown's new hue is not to last,
For brown's no more than red, yellow, and black.
So onward we move: *What do colors mean?*
What's nature's physiological scheme?

When we see red, we see danger: *Stop! Blood!*
Metabolism rises, adrenaline floods—
And so restaurants use red tablecloths
To increase both the appetite and the cost.

Yellow, the quickest color we can see,
Means caution, as with black on a bee,
But yellow's bright and cheerful too, and lends
Light to small and sunless rooms like kitchens.

Healthful orange is the common man's color;
So to make the expensive look cheaper,
Such as with a hotel, they paint it orange,
And put some shiny polish on the door hinge.

Blue invigorates and therefore provides
Extra strength and power; so blue's on our side
When the home team's locker room is painted
In its hue (visitor's was pink—they fainted).

Blue as was said is good, except on food,
For few foods are blue; so in diet mood,
Put a blue light in your kitchen—and lose
Weight avoiding repulsive looking food.

Pink (red tinted with white) debilitates,
Sapping strength and temper, so that is why
It's used in prison cells and locker rooms,
For it calms the most violent inmates.

What of purple? Well, it's mournful, but too,
It's stately, regal, and virginal, new.
Of green, though it's seldom worn, none complain;
And use it in their carpets to stay sane.

The stars are not just white, they scintillate:
Sirius is blue, its companion green;
Betelgeuse, red; many, like Sol, yellow;
Arcturus, orange—all jewels constellate.

Well, as colors go, so then do we, see:
Hues are just differing wavelengths of light
That the brain interprets, in its own right,
For some natural colored necessity.

May I chance upon a land of strange rainbows
Of elfin-hued flowers: red delphiniums,
Black tulips, orange fuchsias, white marigolds,
Bronze grass, and the legendary blue rose.

THE SOLIDARITY OF THE CONCORDANCE

The blend of the coalition grows upon itself,
Striving for the dynamic-balance: of light
And dark, Yin and Yang, and wrong and right.

Reality's not found in separate actions,
But in related events blended in twilight.

The concept of Classicism accentuates
Order and clarity of thought, simplicity,
Restraint, balance, dignity, and
A mistrust of emotion and excess;

However, since it relies on imitation and
The acceptance of objective standards,
It may lack spontaneity and degenerate
Into excessive traditionalism
And empty formalism.

Romanticism embraces an exaltation
Of the feelings, an individualism,
With new modes of imagination,
Of freedom of form, spontaneity,
Self-expression, and subjectivity.

It began, at least in art, music, and literature,
As a revolt against 18th century doctrines
Of restraint, forms and rules, decorum,
Stagnation, and blind tradition;
However, romanticism and classicism
Are now taken as more general terms.

Some exemplars of their contrast are:
Passion as opposed to reason;
The whole against the details;
The Yin facing the Yang;
The right vs. the left side of the brain
"Don't confuse me with emotion"
Or "don't confuse me with facts";
The sails confronting the rudder of the soul.

This epitome may become a battlefield,
Or it may grace a smooth sailing ship.

How easy they are not transformed,
These apparently opposing forces
That may wage war upon the other,
But how tremendous they can be
In the bond of confederacy.

Pure reason, ruling all alone,
Is a force confining and stale;
While passion, unattended,
Is a flame that burns
To its own end.

Poetry is an ideal of the unison:
The right side of the brain
Provides the inspiration;
The left side devises the rhyme.

An utter, absolute classicist
Or romanticist is an extremist!
S/he honors one worthy guest
In the house above the other,
And so loses the love and faith of both.

Witness the average classicist at work,
One who knows little of the humanities,
One who ever works through lunch,
Never having the time to hear of life,
Making every decision by the book
But little from the heart.

Or the total romanticist:
One who can't even hold a job,
Even taking drugs, and losing all control.

The writing of this page—this analysis—
Is rather a classicist undertaking.
But I do not live by the unbending way
And therefore my songbird
Is never imprisoned within.

Perhaps it chooses to be here, classically,
Or perhaps it will, at any time of day,
Burst forth and enjoy the total feeling.

Nor does a long wild night of lovemaking
Mean that you've gone bonkers.

Life is full of spikes of valleys and mountains;
It is only when one can't merge the two
Or at least make jumps between
That one may need some reflection.

How can there be any sort of resolution
Of a dichotomy in which one side
Expresses itself so logically and
The other in emotions and images?

Well, if either one's sails or rudder be broken,
One will soon be dead in the water...

Therefore, the discord and rivalry
Of one's elements must become
Rhythm and sweet melody!

It's not the same for everyone,
But the knowledge of
The 'contrast' itself is the first step...

Therefore, let your blended soul exalt
Your reason to the height of passion,
That it may sing and fly about,
Letting it direct your passion with reason,
That your passion may live and survive
Through its daily death and resurrection, but
In effect ever arising from its own ashes.

Now no one can ever achieve
The ultimate and perfect balance
Between classicism and romanticism,
But for the rare times when in the 'zone',
And indeed, this balancing attempt
Itself smacks of classicism!

And so we all have leanings—
And that's what I mean when I say
My tilt is toward romanticism.

Emotion, slightly favored, rules,
But every so often I do check in
With logic and analytical reason.

Thereby I enjoy the world, mainly,
Because like many of you

I am much impressed by its wonders...

Without perception's deeper depiction,
One finds little that excites—
Not noticing much, as ever in a hurry,
And seldom having the time...

Two other poor relatives
Of classicism and romanticism
Are substance and surface glory.

The romanticist in me likes the veneer
Of the shiny red car or motorcycle
But the classicist in me would like
To know that the vehicle operates well
And even be able to take it much apart,
For that is the very substance.

When I maintain my car or cycle well,
Shine it up, and then speed off
Into the country sunshine
With the wind on my face,
Then I have the best of both worlds!

Now I really don't know all the answers—
I just like to tug at the hem of the garment
In which life's mysterious dualities are clothed.

As ever as in all good marriages,
"The oak tree and the cypress
Grow not in each other's shadow".

People involved in the arts may
Like to listen to music while they work
In order to deactivate the left side of the brain
By giving it something innocuous to focus on.

Personally I often dream up many ideas
While listening to music that moves me deeply,
For then the imaginative power
Of the brain's right hemisphere
Is free and inspired to soar unbounded.

Yes I do lean toward romanticism...
Perhaps it is my nature nurtured
Or perhaps I feel a need to counteract

The overabundance of classicism in the world
Or perhaps because in romanticism there is grandeur,
While in classicism there is but cold logic
And endless analytical thought.

But even with these leanings,
The good romanticists never forgets
That it is classicism that pays the bills
That authorizes the indulgences.

I have some hope that
In any totally classical person,
No matter how stern or dull s/he be,
That one day, somehow, somewhere,
There will come some small measure,
And then the ever-during triumph of jubilation.

Yes, the desire to be orderly and factual
Is a part of the human species,
But there are other yearnings in every person—
The desire to be imaginative and unrestrained in
Expressing personal emotions,
Warmly and freely flowing,

And to take in art, music, literature,
As well as escalate the way one lives a life
From an illuminating flame fed from the self,
A source of lucid experience that
Can usher wisdom and fervency,
As the means to the rounded truth.

Then luckily these may be some of its aspects:
Sentiment, celebration of nature, interest in the past,
A new emphasis on feeling and the senses,
Even actually enjoying melancholy and sadness.

Thence comes love of freedom, mysteries,
Even fascinating figures and heroes,
The allegorical, a delight in whimsy,
The improbable, and the 'impossible',
Of legend, folklore, and mythology,
An awe before the immensity of what is—
The Earth as a friend and
The sky as a warm blanket,
And certainly the uniqueness of the self.

The curious blend never lets one down,
Ever keeping one centered, but ranging.
So extroversion entertains at large,
While love's introversion wins one-on-one.

Intuition and sensing
Can sustain each the other
In a magnificent fusion.

Thinking and feeling combined
Are of an unbeatable synergy,
Of a being coalesced and intermixed.

Sensing the general direction but
Not exactly knowing one's next move
Is of a spontaneous higher 'order'.

Here looms the classical planning of
A magnificently grand adventure,
Whether triumphant or of glorious failure
Always of the superb and the sublime.

Merge these ingredients, until smooth,
This loving mix, mingling and combining,
Soon melding into the 'zone', well integrated,
Stirred, whisked, and folded,
In and out, the commingling
That leads to the harmony of amalgamation's union,
The marriage and the synthesis, the very admixture
Of the concoction of life's ever-during brew.

The parts all sum to the whole flow, so
Life must be more like a mosaic done
Than some focused laser tunnel of sun.

Since few lengthy pleasures are lent to us,
We build a stained-glass window of small ones.

Oh thou soul, dare to live near the edge;
Brave the walk of the line, balancing fun
There between adventure and misfortune—
For the greatest blunder in life is to
Repeatedly fear that you might make one.

Hail! Lord Byron's Golden Mean extends:
Let us have wine, lovers, song, and laughter—

Water, chastity, prayer the day after.
Such we'll alternate the rest of our days—
So on the average we'll make Hereafter!

Wholeness arrives by mixing the suspension:
Classicists drone toward dull perfection,
Romanticists drown in feeling's affection;
Worse, others alternate between extremes—
It's not this nor that but a joined direction.

Harmony then rolls along, round and round:
Each holding within it the seed of the other—
Yin reaches climax, then retreats in Yang's favor,
A cyclic movement of rotational symmetry:
Rounded life is the blend of Yin/Yang together.

The perfect balance may still call upon us:
Edges dissolve when opposites are balanced—
Time and dimensional space are transcended.

Everything joins yet remains as itself,
For what "is not" is as great as what "is".

SOUL FOOD FOR THOUGHT

We bless the 'needed' soul with the holy kiss
Of life, it being this which to replace us with;
Yet what did natural selection ever do, in vain,
Spending so extravagantly on the higher brain?

Well I declare, I see hearts that pump the blood,
And all of the chemistry born of that great flood,
As well as cells all about for everything human,
But where from of the same is thought's acumen?

Because I make of this a mystery as those before
I'll suppose the answer there, that and nothing more,
And say that an invisible soul infuses us, running us,
So that we can know all of that not here before us.

THE FOREVER FIELDS OF REALITY

Michael Faraday introduced
One of the most radical ideas in science.

They thought that he had,
For once, gone too far.

Particles became rather irrelevant,
Being mere spigots through which forces flowed.
The real stuff of reality was the forces flowing,
The particles being only the source.

The burden of reality had shifted,
For the space between particles became primary.
Particles were only the intersection
Of the forces that wove the universe.

Forces create stresses in space,
A superhighway
Of how to get from here to there.

An electron wiggles in the sun,
Tweaking the E/M field;
The ripples travel for 8 minutes
Then tickle an electron in your eye.

You see the light;
Light is a tweak.

Physics has never been the same since.
The field concept became real,
The idea being the same as the thing,
Fudging forever the difference
Between something and nothing;

Yet fields are made of something real,
For they have energy.

Einstein called the fields that be
"A change in the concept of reality...
The most profound and fruitful one
That has come to physics since Newton."

Matter then is simply a place where
Some of the field happens to be concentrated.

Matter travels like a wave in a rope,
But the rope itself does not travel.

The field is not so much
Something in space,
But more as of space.

This is why all particles of a type are identical;
For they are each manifestations
Of their fields everywhere the same.

The field takes on a life of its own,
Even when the object that created it is gone.
The traveling kinks continue;
They propagate endlessly.

Where the vacuum is free of matter
It is not free of field, but filled with it.

Energy and mass are the same stuff,
But it takes a whole lot
Of energy to make mass.

Field is thus the bridge
Between matter and empty space.

Fields can't go away,
As they're part of the
Structure of the vacuum—
When in their quietest possible state
They are the vacuum.

This is about as close to nothing
As anything ever gets.

Forces act on things,
While matter is acted upon;
You can walk through a field,
But you cannot walk through a wall.

Kinks in fields can pile atop one another;
Kinks in matter hold each other at arm's length.

Yet somehow, beneath it all
They are kindred spirits.

Faraday made fields real;
Quantum mechanics made them magic—
And lumpy—the currency of QM.

Everything melts via uncertainty,
As when we try to measure a quantum property.

But this too means that no quantum property
Can ever be zero, for zero is a precise amount,
That is, it is that motion can never cease.

Try to pin down an electron,
Such as putting it in a box,
And it increasingly moves about, ever faster.

It is heads or tails while it is still spinning?
Well it is just a fuzzy 'both' yet neither.

In a way, QM eliminated
The very idea of zero
From the physical world,
As 'nothing' never sleeps,
But is ever up to something.

THE WAN MOON

Darkness drains my life away,
Sickness consumes my spirit;
My mantle is heavy lead;
Life's last glow is upon me;
My eyes are craters gone dim.

Death's ebon form seeks me out;
He covers me with his cloak.
"Come away with me," he says,
As he cools my burning brow;
"I offer you quiet peace."

A sudden strength comes to me,
In my waning crescent wisp.
In night's cold shadow I say,
"Un-hold my soul, Moon Reaper,
I shall fully shine once more!"

THE LOAN SHARK

An unusual track was found in a cloud chamber
That Carl Anderson was using
To watch the trajectories of cosmic rays
Streaming in from space.

The track was like that of an electron—
Except that it curved backwards
Under the influence of a magnetic field.
It was the positron, now used in the PET.

A particle and its antiparticle annihilate,
Giving back in the process the energy it took
To create them in the first place.

Do they live on borrowed time and energy,
A creation near 'ex nihilo' all over the universe?

Can they sneak out of the vacuum
So long as they snuck back in again
Before one has noticed?

"What is the point?"
Thought Richard Fenyman:
"Created and annihilated,
Annihilated and created—
What a waste of time."

They come and go like dreams,
The lighter ones, like electrons,
Popping out more often.

They are the ghosts of the yet unborn.
The road from 'nothing' to something
Goes in both directions;
With enough energy
They can become real.

The so-called 'vacuum' is creative.
The field fluctuates this way and that,
But on average the net energy is 'zero'.

The once melted vacuum fell and froze,
Gaining structure,
Such as when water becomes ice.

THE MEADOWS OF HEAVEN

We of the highest consciousness ever known
And of the most versatile form that's been shown
Reside as consequent beings in this Earthly realm,
Possibly the most fortuitous creatures
That the universe has ever wrought.

In fact,
We are this universe come to life—
Necessarily from a long line
Of 'fortunate accidents'.

It had to be this way, for any universe
In which we could emerge
Would have to be appropriate for us
Or we wouldn't be here to discuss it.

Looking back
We already know ahead of time
That we will discover
The many 'happenings'
That made us possible.

All this we know and expect
Because we are here.

Perhaps in some other 'wheres',
Junkyard universes litter the omniscape,
For they flunked, failed, and miscarried—
A quadrillion trillion universes broken down
For every one that worked to any extent at all.

In some of these forlorn universes
Perhaps the material was inert
And so it just sat there, doing nothing, forever.

In others maybe gravity was insufficient
Or had no natural place to collect particles
And so it thinned out endlessly,
Spreading coldly toward infinity.

In yet others again,
Even those in the same ballpark as ours,
Perhaps the portions weren't quite right.

Although they may have formed a few elements,
They went no further than that for a zillion years.

It would also be that all the possibilities-probabilities
That are of so many imbalances must ever trace back
To the Perfect Imbalance of matter and antimatter,
This no longer seen as improbable.

In our universe the dark chest of wonders
Of Possibility and Probability opened up
In just the just right way:

Naked quarks spewed forth,
Among other things,
And boiled and brewed
In one of the steamiest broths
Ever cooked up.

They somehow simmered and combined
Into the ordinary matter
Of protons and neutrons.

Then quite independently,
By some unknown means,
Dark matter-energy arose as well,
In just the right mix, and, luckily too
Some very long filaments,
Called cosmic strings,
Formed and survived long enough
To be useful as collection agents,
Which were merely imperfections,
As in an unevenly freezing pond—
A kind of a cooling flaw.

None of these happenings were connected,
Except by Potential's destiny,
So, 'fortunately',
The cosmic strings attracted,
By their gravity,
Both dark and ordinary matter,
Which in turn
Attracted even more of the same.

These pearls of embryonic galaxies arose
And were strung along these cosmic necklaces,
As can still be noted today.

So it was
That some almost incidental irregularities,
Frozen out as cosmic anchors,
Were latched onto by matter, both light and dark,
The proportionate portions of which were favorable,
The dark matter dwarfing our ordinary matter
For some reason of a happy 'circumstance'.

'Fortuitously', as well,
Anti-matter, if there ever was any,
Did not fully cancel out the uncle-matter.

The universe could not foresee any of this
In and of itself's fundamental substance(s),
For if it could have
Then we'd only have the larger problem
Of how the foreseer could have been foreseen,
Ad infinitum...

So it could have been like the 'trying out'
Of all possibilities in superposition...
A brute force happening
Of every path gone down.

We know much of the rest of the story
Of how the stars and their supernovae
Created the light and heavy elements
Which combined into molecules,
Which 'auspiciously'
Became able to replicate themselves, as DNA,
And progress to make cells, tissues, and life.

And then there was the luck of oxygen,
A mere waste product of photosynthesis
By bacteria, and later, plants,
That could fill the lungs
As well as build an ozone layer of protection
From the harmful rays of outer space.

Luck on top of luck, good fortune,
And then prosperity...
'Stumbled along' the right path.

Of course all this took many billions of years—
And it is of course this long 'yardstick'
That baffles the mind and sticks in the throat,

But demonstrates the long time lag needed
To produce even the tiniest of advances.

It bears all the hallmarks
Of 'randomness' at work,
Although quite probable
If Potential had it all 'worked out'.

Dinosaurs roamed the Earth
For over two hundred million years—
Imagine the length of that time.

They were supreme and invincible—
The kings of all the Earth 'forever',
On land, sea, and even in the air—
Heading towards forevermore and beyond,
But...
Dame Fortune once again intervened
When the asteroids or some such catastrophe
Finished off the dinosaurs,
As well as 90% of the existing species.

This 'random' event left a vacuum
In which newer species could thrive.

Proto-man gave way to near-man
And thence to us, eventually,
When two 'monkey' chromosomes fused together,
Making 'us' incompatible with the chimps
And so our ancestors then
Truly descended from the trees!

'You' were once a lucky shrew, darting all about,
But then attached to a favorable evolutionary line...
Every single one of your forbears on both sides
Being attractive enough to locate a loving mate,
And they fortunately had the good health to celebrate!

We came to need no specialized niches,
Since we could adapt to any terrain,
Having brains that could learn much more
After birth than instinct could bestow before.

Our higher consciousness
Was the crowning glory—
We had won the human race—

The be all and end all; the grand prize
Of the universal lottery.

So there is nothing more,
Aside from our own progress
To be and learn.
This is it!
That's all there is.

DNA remembers every step of our evolution—
And you can see this in 'fast' motion
When embryos form simply in the liquid womb,
Replicate, and then grow cells
That diversify into a human being
After going through some nonhuman stages.

Thus four billion years compresses into
The nine months of pregnancy.

So then hail and good fortune,
Fine fellows and ladies,
And welcome all of you
To the Meadows of Heaven—
The highest point of all being,
Although we are surely
Still in our infancy.

The path "chosen" by Potential ends here,
With our consciousness.

There were many pockets of universes,
And it is was this very one
That could sing our verses.

The further design
And the role of mankind
Is now in our hands.

We were borne here upon the shoulders
Of so many who have long since come and gone,
All of them advancing the cause,
Over eons of wiles—so here we are.

Fare thee always well, fine friends,
For we are some of
The luckiest sons and daughters of being

In a rare universe well done.

Celebrate; live; be,
For everyone dies,
But not everyone lives.

THE KARMA OF THE BARKING DOGMA

Some Hindus, Buddhists, Christians and Jews
Wondered what stories they should choose.

Even thought they'd already so many chosen,
They just didn't want to keep notions so frozen;
So they met to merge the postulations into one,
Thinking that this might be a whole lot of fun.

"In our hypothesis, there is just the only One."
"Well, our conception is a multitude of many Some."
"Well, we'll part way meet: there's only the Holy One"
"Nah, the odds of that are over three million to one!"

"Buddha of us was one, so of Gods there are none;
A human above all that now's not seen by the sun!"
"Humph! Holy Jesus of our one God was His son!
He lit mankind's darkness with light of the Sun!"

"No, Jewish Jesus was not of any nature Divine,
But was just a mere man much ahead of his time.
This you all should know, being there at the time.
Look at our history singing old biblical rhymes."

"All is not real, so what is this great big fuss?
Retreat back to where it's all at to slow the rush."
"Oh God's universe and creatures are so real
And that is why we're making this very big deal."

"In the afterlife, we in Hell or Heaven reside."
"Not so fast, for in between these realms we lie,
And if you in this testing life don't do so well,
You'll have so many subhuman tales to tell."

Reason arrived: "Possibility reigned way then back
'Before'; there's nothing even holy about all that.
'Tis all made up, those many fabrications made,
So just let it all be, for this is what existence bade.

THE END OF THE EARTH

The Asphodel sustains the Dis dwellers,
Where they rest beyond that fatal river—
There the wretched shades drink forgetfulness,
And to oblivion sink without distress.

Charon was withered, wan, and skeletal,
Although eternally grateful for his immortal life
And steady job of ferrying the dead across the river Styx,
In their transition from life to death to forgetfulness.

Fireweed grows from Hell's sulfurous embers,
As does Purple Loosestrife—dead men's fingers;
But wildflower air revives the dead—and then
Those happy souls can thrive on Earth again.

As Earth was the only planet he'd come across
With such promising higher life forms,
Charon had grown rather fond of its inhabitants,
Even though he only saw but the worst of them;
But even from that he could extrapolate
To the qualities of the best.

Charon did his job well, professionally,
Although it was ever so dreary,
With the endless darkness of wasted lives
And the grim and gloomy skies all around.

The land always had
That same gray and leaden feel.
He ferried on, though,
For his own life was precious to him.

The soon-to-be really really dead never said much,
For what was there to tell after an empty life
That had often turned to deep regret.

Charon did not prompt them for information,
For this was not the thing to do
At the time of their passing,
So he was always most
Courteous and kind to them,
Even to the most evil of the darkest,
Doing his task as well as he could.

It was not that Charon was afraid that
His undersized master of the underworld,
Pluto, might be watching,
But that he had the extreme clarity
To duly serve the task at hand—
A testament to his character.

Charon had been quite alarmed lately—
What with the numbers of the hellish-souls-to-be
Climbing into the millions in such a short time,
But he had been through this kind of rush before
With the doomed and damned of other planets
That had been consumed by their suns
Or had undergone other such catastrophes.

He just used larger boats,
And patiently took his time,
For he had all of Eternity.

Of course,
Charon could and did feel deep sadness,
But he didn't show it outwardly,
Even when the numbers from Earth
Increased a thousand-fold again.

A few of the now billions of depressed Earthling souls
Had enough energy left to mumble a few words,
And so he was able to glean from them
The latest happenings on Earth.

In 2015, the predicted exponential surge
Of melting ice from global warming
Had quickly inundated all of the coastal cities,
Many of them large centers
Of population and commerce.

Everyone who could possibly make it
Had to retreat inland,
Creating the largest mass exodus in history.

As the heat rose to unbearable levels,
Many had begun living in their basements,
As the Earth's infrastructure
Began its eventual collapse.

Millions eventually headed north
Towards Canada and Siberia,
But had to retreat when the ice caps totally melted
And formed the great Ocean of the North;
Most of them did not make it.

No one but the ignored physicist mathematicians
Had predicted that the end
Could come into sight so quickly.

Then came the dreaded polar shift
That made global warming seem but a small note
Compared to this new and darker symphony.

The Earth was thrashed with storms
The likes of which it had never seen;
Electricity went out completely all over the world,
But for a few nuclear powered areas that didn't last.

No one could drive very far,
Even on their last tank of gas,
For the roads had melted,
Along with the tires of the vehicles,
And if the vehicles stopped
They'd find themselves mired
In the meltdown of the asphalt.

Food would no longer grow very well,
Even in once lush gardens,
In the amounts that were needed,
And, as the heat rose further,
Into the 140s, plant growth ceased altogether,
Although a new but rare
And expensive form of food pill
Extended life for some of the rich,
For a short while.

Charon had of course,
Seen much of this kind of thing before,
From the many other solar systems
And galaxies on which life had formed.

Earthlings seemed to have
A special charm and hope
Above and beyond the other alien races.

So he rowed and ferried
And deposited them on the far shore,
His job and life forever continuing
In a place with no color,
No joy, and no future—
On the shore of the land
On the edge of oblivion.

Charon had depths of compassion,
But many passengers might
Have thought him stoic,
Although they were mostly
Beyond this capability.

A sign on the opposite shore said:

Abandon Hope All Ye Who Enter Here

Billions more arrived
In the gray land all too soon
And Charon learned that
Either madness or desperation on Earth
Had caused a nuclear winter all over the planet,
Bringing on a deep freeze that few could escape.

Perhaps they were trying
To combat the ultimate heat,
Which would have been
But a cool breeze in Hell.

The polar shift had greatly
Added to the deep freeze.

A few of Charon's still speaking
But chilled customers
Even expressed a longing
For the legendary warmth of Hades.

Charon, stalwart and reliable, rowed on steadily,
Ever steeling himself to the misery.

Finally the masses slowed and dwindled
To a few dribs and drabs over a few years
And then there was no one for several years.

A lone man appeared on the shore near the ferry dock,

And Charon readily approached the man,
Something he had never done before.

They had a long and hearty talk,
For the man was animated
And not at all like any of the other wretched souls.

"How is it," inquired Charon,
"That you are full of life and seem to be a good man
But have been sent here?"

"I am not a bad person in any way," the man replied.
"Actually, I just spent some time in Heaven.
I found out there that my sweetheart
Was sent here to you,
For she was a suicide
And so was destined here;
However, I had promised
To be with her forever,
So I chose this place
Over Heaven out of my love for her."

"Extraordinary," exclaimed Charon.
"I knew the Earth had
A few good men and women;
I've not seen very many clues
Of that elsewhere in the universe.
Did you colonize space—
Will your species continue and flourish
After your Earth bids farewell?"

"I'm afraid not," replied the man,
For too many needless wars intervened
And this greatly delayed our space program."

"A shame," said Charon,
But is there any hope left on Earth,
I mean, are there any others still about?"

"I am the last," the man answered slowly.

The first tear of Charon's long life
Rolled down his cheek;
Nothing had ever made him cry before:
Nothing had ever made him weep.
(Rewritten from Lord Dunsany's brief sketch)

HIGHER CONSCIOUSNESS

The three lower consciousnesses that are
Obsessed with the securing of objects,
With the chasing of sensations, power
And control will never ever be enough.

There are No actions of people that can
Justify our becoming irritable
Angry, fearful, jealous or anxious if
We give them our unconditional love.

If we don't accept the unacceptable,
Then we lower our level of consciousness;
Our response will mirror their uptightness—
Which can spread the bad moods onto others.

Conscious Awareness, which can but witness,
Is a safe haven from which to observe
The drama of our lives playing in our minds,
Granting us a sobering distance from it.

From a safe subjective place that's free of fear,
Our soul, our conscious awareness, can witness
The strange thoughts and emotions that surface
On the mind, sent by the subconscious brain.

Putting ourselves in the place of others
When hurtful things are done to us
Expands our consciousness, compassion, and love,
Since we can come to know why they did it.

When we converse with ourselves it is our
Higher Consciousness—our Conscious Awareness
Or I that questions our lower consciousness
Impulses toward securing, sensation, and power.

Seeing the big picture of life and its stages
And connections lets one not get annoyed, say,
At being cut off in traffic, for they
May be old, learning, lost, growing, or angry.

Putting the needs of others ahead of
Our own produces the byproduct of
Happiness and reduces stress, for we
No longer have unrealistic expectations.

There's No Life in the dead past, just history,
Nor in the imagined future, a mystery,
But in the here and now life just arrives;
Its a gift—that's why its called the present.

WORDS VERSUS ILLUSTRATION

The writer's pen stood forth, being first,
Instructing the artist's stylus
To illustrate the words of the epic,
Noting that a picture was worth a thousand words.

"Perhaps we don't even need the words",
Retorted the artist's stylus,
"As I am worth so many".

"Well," replied the writer's pen,
"It's true that many people now refuse
To read books without lots of pictures in them."

"How sad, for I guess some words
Are needed to round out the tale."

"True, for the two sides of the brain
Can then combine in unity."

"Or I could draw the pictures first
And then you could write the words."

"It could be like that sometimes, I suppose."

"OK, shake; it's a deal either way,
For we need each other."

THE NEAR DECLINE OF PHYSICS
DUE TO ITS UNDRESSED TERMS

The quarks, those constituents of the orgy
Playfully bound within the nucleons' chamber
Are named *up, down, strange, charm, bottom and top,*
The last two once being called *beauty,* and *truth;*

However, when just one of a type was contained
It became referred to, say, as a naked beauty,
And thus nude tops & bottoms their charms revealed—
To ever be in closeness binding, and bonding,

So they even tried just *u, d, s, c, b,* and *t*
To prevent some ultimate collapse of physics,
But the truth of the flavors beneath the veils
Remained as the sheerest vision preferred.

So we have these vibrant dancing ladies:
The naked heavyweight top, charming up,
Down, the strange beauty of the raw truth,
And a bare bottom just around and behind.

They gyrate, spinning their charms, twirling
In the universal dance of stunning motion,
The polarity sometimes reversed,
Whirling, their bottoms up and tops down.

And then there are Eden's many colors,
In this flower garden filled with flavors,
Such as red bottom beauties, blue tops,
And magenta undulations unstopped.

Gluons are the bees of the flower beds,
Carrying pollen back and forth to bond
The many relationships that make
This loved world go 'round as reality.

Eyed in views that probe the fundamental,
Quarks strangely swirl in and out of sight,
Pulsing, throbbing with elemental delight,
Back and forth—the love-made life of eternity.

These attractions in the altogether denuded
In the buff became the strong force manifest
That these mother-nature-naked terms exposed

To denote the stark beauty of truth uncovered.

THE ENTRANCING DANCING

They were all dancing within love's treasure vault
Within the framework of the broadening thought,
The lights pulsing and the waves reverberating,
Where the good times had become everlasting.

Tribal primal field currents were raging
From speakers of the energy matrix pounding;
They whirled and twirled as loving gestalts
Of sentient consciousness knowing no halt.

There were rhythms of constant contraction
And expansions of bosom-energy projections
Converted to scalar waves of blinking attraction,
As fission and fusion beckoned the connections...

Ever forming in this Omni-sound emporium,
Where tone waves vibrated in waves of creation.

"THREE QUARKS FOR MUSTER MARK"

Naked quarks would really love to go wild and dance,
But there's only a finite amount of energy and chance;
So they would spiral out of control,
Having quite a blast!

Such they've been confined within the proton—
To last.

They're made bottoms-up;
Can we see them tops-down, a go-go?

No, for the quantum censor protects the charm show,
Their strange beauty and flavor bound up and down,
For the proton is much immune to disturbance around.

CHARMS

A new kind of microscope
That works via gravitational waves

Has revealed the actual interior
Of a quark for the first time.

The charming beauty
Of the ultimate truth
Is that ladies are
In charge of the universe!

WE ARE MOST FREE WHEN
WE ARE ASYMPTOTICALLY CO-JOINED

The strong family unit, as the three quarks,
Is bonded by the power of its grouping,
But, loses identity if the home breaks—
Other pairs soon forming after divorcing.

Or comes the prison of solitude,
Chained to isolation with fortitude,
Floating, lost, without effects of affects,
Losing the identity conferred by others.

Within the proton, gentleness becomes strength,
For the members are free to explore at length,
Never smothering, but building unity,
The unit's direction adding to the one.

The strong force grows weaker near the quarks,
And so we may observe them someday,
Shining in their primordial glory—
The beginning of all things composite.

Identity is not lost in the co-joining—
True loves don't crowd the hearts of the others,
But, rather, look outward, in the same direction,
Close, joined, but don't block the others' section.

It is a seeming arithmetic violation,
That in summation we become greater;
We don't merge, having supported freedom,
Yet still share the same good vibrations.

Love matures when partners let it flow beyond,
Free to wend its way to places dear and fond.
Love's butterfly prospers when winds blow free;
Unconditional love never binds—it bonds.

WISDOM
(wise-dom)

It's the superior judgment, understanding,
And application that is based
On both knowledge and experience—
Far surpassing erudition; the quality of being wise.
The antonym is "folly".

It goes so deep that one may even
Easily ignore one's own (conditioned) thoughts
Which arise that are unknowable beliefs
Falsely identified as truth and fact
(A second level view: beliefs about beliefs, sort of).

One who has it may be be called a Wiz
(No relation to the magic of a wizard).
Learning feeds it to deal with the quiz.

Some run into the walls of life,
Time and time again, ever bashed and injured,
But never ever learning.

"Wishes" seen but only through one's own eyes.
"Say" that they ought not to,
That they shouldn't; but,
Wisdom notes that they still do, the reality—
That they can't, they don't, and they won't.

Such is the human condition for some
That they may be immune to learning,
The curse that prevents the will
From becoming wider and having more choices.

Yet the ultimate vision remains available
For the rest and one day the "some"
May be swept up into its sum.

TRUE COLORS

We are the Eternal Smile of Being,
The Joy of the Universe's Creation!

In us the Cosmos has come alive,
And has evolved into our consciousness,
From primordial matter and energy.

We have arrived! We are the Cosmos itself.
We are the Universe—life from Stardust!

We live but for one of Eternity's heartbeats,
Borrowing Life from Death for just a while.
All that we are we owe to Time, Death, and Stars.

Truly from the Stars cometh our help;
The Stars are the creators of atomic matter.
Within a Star's heart, matter transforms itself
And gives off energy; this is why the Stars shine!

Death is the ultimate evaluator
And the director of all evolutionary progress.
Death selects the wise from the silly;
Death chooses the useful from the useless,
But it takes Time.

It is this long yardstick that sticks in the throat.

For what seemed like Forever,
Our sleepless spirits have waited
To catch light, life, and delight
From Heaven's smile.
Finally, we are so lucky and we live.

We stand atop the pinnacle of Nature's tireless toil,
Which has at last brought forth our souls
From that black and endless eternal deep.
What a joy to Be!

In what far and fiery depths of space
Burnt the fire of your Spirit?
In what distant star was born the gleam in your eye?

Know it well, for one day Death will ask you
"What did you do all of your life?"

But for now we are alive.
Our mind and senses interpret and disperse
The base Reality into the colors and sensations
Of the phenomenal world.

We can become either rainbows or ugly stains!
Our minds, like Shelley's prisms of many-colored glass,
Strain the white Radiance of Eternity,
Into our life—until Death tramples us—
And back we go to stardust,
After relentless time has wasted us away.

Yes, our creators of Time, Death, and Stardust
Must also write our epitaph;
They devour us in order to return
That life-dream which was lent to us.

But here we are now, and perhaps we come to know
That the simpler things in life are still the best:
A glass of water from the well in the morning;
To love, laugh, and sing with family and friends.

And so we live out our lives with honor and love,
Kindness and generosity—these are our true colors.
Life for the sake of life! Good for good's sake!
Enjoying everyone and everything and every season.

Many think that they are more important
Than they really are, that they deserve some reward—
Of a divine destiny in Heaven where their every whim,
Wish, and fancy can be fulfilled for all of time.

Well, to me, such endless satisfaction and pleasure
Sounds really rather prideful, wishful, even decadent.
The ultimate humility is for us to realize
That we are just electrochemical organisms.

Are we quite lucky and fancy organisms? Oh, yes.
Are we specially created by a Master? Oh, no.
We are the embodiment of the Cosmos
And are ever the results of natural laws.

Death will be forever, but man,
With his exaggerated view of self-importance,
And not wishing to see a final end

To his glorious life—and I can hardly blame him—
Desperately grasps for immortality's promise.

For me I will continue to catch life's joy and smile
And will bathe in the light of its constant sunrise.

On my last night on this Earth,
I will not be haunted by regret
When the Sleep of Death comes
To take me to Corruption's dim dwelling place—
For I will know that I lived for color and smile.

And what of the Stars?

They remain, as Eternity's Love-lamps,
Representing our good works and deeds,
Which even the fathomless night cannot quench.

Perhaps one day, at the end of forever,
The Stars too will die and grow cold when
Time conquers all; but as long as they live
They will shine and radiate the hues
That paint the colors of our ashes
Reborn again on the phoenix wings of Time.

HOME SWEET HOME

Some people drive treacherous roads
Up to a mountain top
When it is snowing so they can ski,
Paying a lot of money
For the privilege of breaking their bones.

They must have poor mental health,
For they even put slippery wax on their skis
So they can go even faster
Down the mountainside
That is already slippery
And full of snow and ice and trees.

These humans have not evolved much
Since the time of the Woolly Mammoth.

WHAT IS MAN (AND WO-MAN)
BUT SAPIENS SUPREME

From what beastly heart springs our zest?
Of what searching eye became our sight?
What sound in the bushes let us hear?
What dark past haunts but helps us be?

Across what ink black river did we have to swim?
To what ends at length did we search for food?
In what deep entangled forest were we bred?
Of what stars did we shine of their stead?

Oh Man! What a piece of work—the mind;
What noble deeds done and undone in kind.
What Rube Goldberg inventions heaped upon—
In the layers of brains the mind is made upon.

What is this sapiens mammal animal,
But of some slime and of brutish law!
So, let's 'neglect' this state of affairing,
On the grounds that it is unappealing.

So then...

We are spun of the Eternal Golden Braid,
Those windings of Truth, Love, and Beauty made
From the Goodness of Purity Immortal—
The Theory of Everything's singular portal.

What is Man but the special chosen species
For which all the plants grow and the waters reach,
For which the Earth turns 'round, and orbits
A sunny furnace, spreading Love's energy,
Enabling us to thrive above any and all creation.

What is Man but the only bloom for which all
The 13.7 billions years of evolution and love
Have occurred, in a predetermined random yeast,
To form and flower such a vainglorious beast.

It's ever on forever's edge that we meet our destiny,
That in our temporary parentheses of Eternity
We would flourish for just this moment, bidden,
As the blossoms of Perfection's Flower Garden.

A hundred trillion stars and countless shores
Were built to light our universal nights explored;
Forty million other lower species too, the All-Might
Placed about our world, merely for our delight.

Our names are Writ Large on the Heaven's marquee,
In the supernovae stardust showered from Thee.
From Nothing not You came, but of a naught
Our own universe was made and ever wrought.

A starring role we play in this reality show,
Every atom spinning fine just for us to know,
Our ancestors rising/falling for us to stand upon,
Oh man! They lived and died for our lone promise!

Every shaft of light shines with us in mind;
Thus it beams forth our beginning and our end—
In and of God's hidden and Heavenly Shrine.
Oh life! We cherish being, that of yours and mine.

We do so much deserve reward beyond this role—
And so it is that one's immortal spirit-soul,
That angelic vapour that drives a living being,
Shall go forth to glory on, beyond this scene.

We are not merely some mammally organic luck,
But purposely evolved on this planet, near a star,
In that intended long and winding mindless 'birth'
Of slowly drifting time, dust, and selection by death
That ever sifted the best from the rest: Sapiens!

(Now why is the soul so 'true' and so far with it faith goes?
It is only because one so much wishes it to be what knows.)

Our instruments detect what our senses cannot,
Of the whole electromagnetic spectrum,
Of odours and molecules beyond smell and sight,
Of stars far away and even way back in time,
All because scientists exalt in mystery.

Human introspection and sensation, alone,
Without being informed by the science available,
Is captive to the tales of its second story,
Not knowing the neurological first storey.

So it goes on to declare wishes, beliefs and
'Truth', deepening the wiring upon each visit.
And they might then layer more dogma thereupon,
Till they've a far ranging scheme for life's wonderland.

Things that haven't been established can't be addressed,
For they're 'invisible', such as evil spirits,
So really, a belief in the stated unknown,
'Faith' as defined, can't even be known, much less shown.

Thus to feelings, senses, desires, and sensations,
And claims, we can't trust, true, even their wishing point.
Mysteries shrink away, at an alarming pace,
Now-a-times, and it's hard to keep up with the race.

As for humans, true, we, and our matter that's bright,
Seem to be an afterthought of the Cosmic scheme:
We glow-surf on informational waves of light,
A tiny minority in the grand regime.

Science discovers the truth deep within everywhere;
Religion just makes for begged and bigger questions;
Philosophers just sit around in their soft chairs;
Evolution explains how we mammals got somewheres.

'You' were once a lucky shrew, darting all about,
But then attached to a favorable evolutionary line...
Every single one of your forbears on both sides
Being attractive enough to locate a loving mate,
And they fortunately had the good health to celebrate!

THE MUSIC OF THE SPHERES—MOONLIGHT SONATA

Memory's ideas recall the last heard tone,
Sensation savors what is presently known,
Imagination anticipates coming sounds
The delight is such that none could produce alone.

The music of the spring was in the breeze,
A prelude borne by the airy musicians
Of the trees: the evening calls of the birds
That opened for the cosmic symphony.

The Music of the Spheres played in the park
That night—flung down by our Father, the Sky,
Through the soft night, to our Mother, the Earth,
Then to us, their audience and progeny.

The planets joined in a concert to the
Merrie Monthe of Maie, arrayed as follows:
There was Venusia, the Bringer of Peace,
Singing side by side with warring Marsius.

Flitting about was the wingéd Mercuria,
The speedy messenger who conducted
The orchestra, melting all of us who
Were touched by her wand of burning desire.

And mighty Zeus, was there, full to the brim
With the jollity of the fat man's belly.
By Jove, came Saturnus, so very gray
With age, lumbering into the party.

Thence sat Urania—the magician, and
The old sea captain—King Nep, the mystic,
But not Pluto; he was downsized, no more
One of the harmonics—an underworld!

Jupiter's music was round and robust,
While Saturn's boomed with sounds of grandeur
And the old venerable melodies;
But Mercury soon picked up the pace.

Next flowed the serene love songs of Venus,
Followed inexorably by Martial marches.
Now was the time for Urania's magic—
She played musical jokes and surprises.

At last, their music came to mesh as one,
And our wanderers of the night floated
Away on the haunting, mystical strains
Of King Nep's tune, into the May Flower moon.

Now we're touched, so touched by the starlight,
Afraid that we'll ne'er be the same again.
Can you sense the euphony of the spheres?
Can you fathom the theory of everything?

RAINBOW

Toward the end of a sunny day,
A storm came and washed away,
And the sunset clouds, being glad,
Held a party for the returning lad.

The sun then peeked and soft shone
Into the mist of the departing squall,
Its light split into particolors lone,
Separating, each from the ALL—

A bouquet of colored rays
Swirled into sight,
And promised good weather
For the rest of the night.

The rainbow lit up the east,
As long we attended the feast
Of both the east and the west,
Till into darkness we descended blest.

The stars guided our homeward flight
By shining their jeweled lights
Of ruby, emerald, and sapphire
In living colors of blazing fire.

NOW AND ZEN

Everything that is part of us—
Our cells, tissues, organs and organ systems—
Has come about over billions of years
Because it proved successful
In the great survival stakes
During our perilous evolutionary
Descent (ascent) with modification.

The brain, being no exception,
Evolved in part
To allow a creature to learn
From what happens in its life,
To retain key elements that
Could influence future actions.

We are geared for self-preservation.
We will do anything to avoid facing the possibility
That who we are now cannot continue.

We ourselves are mainly the cause
That we are interested in.
The self is preoccupied with staying alive,
Which is why our species is still around today.

It is a prime biological function to be afraid of death,
And so the self as thus contrived
Is able to fully play its crucial survival role.

We want to equip our brain with a soul
That offers us an escape when the brain dies
Since the self cannot come to terms
With its own extinction.

From a subjective standpoint,
We are all born equal and undifferentiated
(Before that, 'we' were dead),
But as mature selves we make a distinction
Between the individual and the surroundings.

Still the brain keeps changing throughout life
In a pattern of the shifting flux of its neurons;
We gain and lose memories and feelings,
Essentially creating a new person over and over again.

The self is thus not so rock solid as it seems.
These moment-to-moment changes differ from death
Only in degree. In essence, they are identical,
Although at the opposite ends of the spectrum.

So we are not static things.
Other neural networks will come to be in other,
Future people, albeit with an "amnesia"
Of what went on before in
The brains of the previous others.

Why should we be happy about this?

We never can be because the 'I' cannot operate
Outside of its own boundaries.
The only viable alternative is to think of a way
In which it is possible to ever continue on.

What will it be like to be a part
Of someone else after we die,
With our own particular
Narrative of life cast aside?

That is the 'zen'
Of now and then and when.

THE GOLDEN STREAM

In 1865, Hennig Brand thought that gold
Could be distilled from human urine, old,
Perhaps noting a similarity in color,
So he kept fifty buckets in his cellar.

By some method he converted urine
Into a noxious paste of some kind,
Then into some translucent waxy substance,
But so far there was no gold, and none hence.

However, after a time the substance began to glow,
And when exposed to the air burst as an inferno.
The substance soon became known as phosphorus,
But was too costly to make its business prosperous,

For an ounce of the flaming stuff sold
For way more than the price of gold!

BENEATH, BELOW, AND FURTHER

In succession due does the large give way and rule
To the ever smaller, the tiny, the minuscule,
And onto the negligibly insufficient 'awol'
Of not really much of anything there at all.

Yet it was at this bottom herefrom that the all
Of the upward progression began its call,
And so here the answer lies to the sprawl,
At the boundary where nature wrote its scrawl
Of existence upon the non, and back and forth,
A place not necessarily like that we think it is,
A lawless, formless realm that's ever been the quiz.

Stability too has decreased woefully,
Melting within our descending journey,
And so we must meet the perfect instability
Of the potentially perfect symmetry that cannot be,
For not only is it that everything must leak
But that there can be not even one more antique
Of a controlling factor lurking about,
For of anything else we've totally run out.

Here then the pulsations and the throbbings
Of the so-called vacuum that must ever swing
Between here and there, ever averaging to nothing
In its rise and fall, alternating here and varying.

Here Eternity and his elemental fellow rhymes
Of Anything and Everything bide their times,
Of which they have and always had continually
All of the time of everlasting perpetuity,
And so then if one waits long enough,
Which is but an instant in Forever's trough,
Say for a months of Sundays in donkey's years,
Then not only do the rarest of events come to pass,
But eventually so do all things possible that can last.

THE FOREST OF ORIGINAL GROWTH

What would it be like to stumble across lands
That no one else had ever been to,
And how could you know that?

After reading Sir Conan Doyle's 'Lost World'
About dinosaurs on a sealed off plateau of a volcano,
I wondered if there were any more undiscovered places
That the paths least followed could lead me to.

So while at the Earth Summit in Rio last month
I forayed into the uncharted regions of Brazil,
Having chosen from a map the remotest area.

After various vaccinations and preparations,
I trucked my one-man helicopter
To the last way station,
Loaded the extra gas tanks onto it
And flew into the heart of darkness,
Eventually gliding down onto a grassy field
Just as the gas ran out.

From here I walked for tens and tens of miles,
Always taking the most difficult path
Whenever there was a choice,
For this would insure that I could end up
In some totally unvisited region
That was near impossible
Or hard-to-get-at in any usual way.

After hundreds of these
"Improbable" path choices
I suddenly came across acres
Of Lady's Slippers flowers.

These are very rare flowers
That usually appear in small bunches,
Growing only in conjunction with a rare fungus,
And even so usually get picked,
But there were millions of them.

After taking one last really
Difficult choice of a path,
I discovered entire fields of other flowers
Long thought to be extinct.

Some were Eve's Blossoms,
Which not been seen
For thousands of years,
Historically valued for
Their life extending elixir,
As well as the original, lost,
Strain of Pearly Everlasting,
The flower that never dies,
And so I suspected
That I might be in virgin territory.

How would I know?
Well for one there were no paths left,
For even animals and their hunters
Had either long left or had never even been here.

Also the flower colors were not like any
That I had ever seen before,
Not new colors, mind you, but just, well,
Colors of different intensities and hues
That were not thought to exist in nature.

I saw true-blue roses, legendary no more.

I had chanced upon a land
Of strange rainbows of elfin-hued flowers:
Red Delphiniums, Black Tulips,
Orange Fuchsias, White Marigolds, Bronze grass,
Yellow Violets, and even Adam's Apple,
Now growing from the ground!

Was this the original forest—
The Garden of Eden?
Was I the first to return
To this legendary paradise?

And then I knew that
It was indeed the Garden,
For there, right in front of me,
Was a field of thousands of undisturbed
Golden nuggets on the forest floor.

Surely no one had ever been here,
At least not for a long, long time.

I reached up and put the apple back on the Tree.

SHADES OF THE OTHERWORLD

Could there be more to this world—
Those of the undrawn shades unfurled?

Is there a universe alongside this bright zone,
A parallel, twilight world overlapping our own?

Are there shadow beings all about us,
That we can only perceive as blankness?

They'd be made of but the dark matter,
Yet lively with their own kind of chatter,
These shades flowing right on through us—
We the lighted "plus" to their dark "minus".

These pale shadows of our attendants,
Are not as us of light's extent,
But are as black clouds of a coal sack;
Nay, they're not even dark or black,
But are of an invisible bivouac.

Dark matter and its shadows traverse
The bulk of our missing-mass universe.

The shades of evening draw us on—
We must look to the past, upon the first eon.

Two distinct families of matter
Were created in the Big Freeze batter,
Just those two that did then so accrue
When they were frozen out of the primordial stew
As the fetal universe was cooling,
When the hearty gruel was ungrueling.

The normal universe and the shadow universe
Can interpenetrate, neither averse
(Or even "adverse" to rhyme the verse)
Nor of coerce; they just cannot interact,
As they have no contract.

If the shadow universe was richly sown
It could have evolved along with our own.

Shadow planets could form
Around shadow stars as norms

And become populated with swarms
Of those shadow beings lukewarm.

They would be invisible specters, unseen phantoms,
Unobserved presences, indiscernible apparitions,
Imperceptible wraiths, unnoticed spirits, magic places,
Inconspicuous spooks, and hidden traces...

But first we must ask what makes a universe,
Such as ours, the one in which we immerse.

It is the forces that count for everything,
Matter being but a secondary singing,
For atoms exert forces through space,
Especially of the electromagnetic race;
So then it is forces that disburse
The currency of a rich universe.

This is why we don't fall through a chair—
That mostly empty space of "thin air"
When we decide to sit down there.

Space is a kind of a large-scale limitation
Of an underlying discrete network of connections.

Atoms would not even know at all
That their companions existed, with no call,
Without the push or pull of the forces' thrall,
For then they themselves would be as pall
As some ghosts passing through a wall.

The four forces hold our world together
In its diversity of shape, structure, form, and color.

Some forms of our matter don't feel
All of the four forces as real:
Neutrons have no electric charge
And so they don't "care", Marge,
About that e/m force at large.

Suppose some form of matter didn't feel
Any of the four forces that became real?

Dark Matter doesn't appear to discourse,
Not having the resource of its own special forces
To bind it together; no packhorses.

All it can feel is the force of gravity,
And perhaps the weak force's changeability—
Which is for decay and not stability;
In fact, both forces are weak, a pravity.

You cannot hold a person-size lump
Of matter together with just gravity's slump;
So then no interesting lumps can form
In the dark universe, not even unicorns.

Even making a star or a planet
Is difficult with just gravity alone working on it,
For the electromagnetic force is crucial
To slowing any of the material
Down enough to hold it in one place;
So then there can be no shadow race...

...No veiled hints, obscured suggestions,
Unknown impressions, out of sight suspicions,
Nor any supposed tinges, shimmering glimmers,
Resembling semblances, or ghostly whispers.

What has no light is but a dark shade,
With no creatures therein made.
So dark matter is not a source for being;
'Tis but a very large matter to us unseeing.

And yet is it we who are the outsiders,
Our luminous bubbles of foam the riders,
The stars, planets, and us the striders—
On the vast ocean of dark matters much wider.

We were an "afterthought",
With no forethought,
Although perhaps made possible, nonetheless,
By the dark matter—since it was oblivious
To much of the great primeval blast,
It forming filaments that could last,
Attracting our regular matter
That was everywhere splattered,
Into the pearls of the galaxies
Strung along like cosmic necklaces.

Far from being the *Magnificat,*
We are more insignificant

Than we ever imagined,
For whatever is our measly count,
Compared to dark matter and dark energy
We're but a kind of pollution, irrelevant, really.

BILLIARDS

All of pool that they say is so true,
For my father bought me a table too.
Being in the right spot at the right time
Is one's good fortune created on the dime.

After homework was all done we'd make the run
In billiards playing banks and carom fun.
We'd have to hug the girls to teach them how,
Then even demonstrated 'kissing', wow!

After school we too played some pool
In the smoky de, where gambling fools
Bet their lunch money time and time again,
Where too some girls mixed among the men.

I won an Army eight-ball contest,
Winning a carton of useless cigarettes,
So I gave them away, for the play's the prize,
Keeping me here and away from paradise.

I haven't played much in forty years,
Nor have I drunk even that many beers,
And some long shots I still had to make,
But the short easy steps were best to take.

I turned to tennis, the theme much the same,
Though the bounce was but upon the ground:
Don't look up too soon from the knowing aim
And follow through—works too for other games.

Never played much golf but did sometimes,
Driving was all right, but nothing really fine,
But upon the putting green my eye was right,
Sinking putts like billiard balls, left and right.

THE KNOWING

Into this Universe, and why, not knowing,
Nor whence, like water willy-nilly flowing:
And out of it, as wind along the waste,
I knew not whither, willy-nilly blowing...

Now I'm knowing, that out of this muddle,
Indeed, it's the chaos that frees me to be,
For it's all of disorder in disarray—

An ultimate disorganized confusion,
Whence all sprung, banged, and exploded,
With no hint or trace of order, law, or plan;

'Twas mayhem, bedlam, and pandemonium,
Wreaking havoc upon the turmoil of a tumult,
Heaping high upon, a commotion of disruption,
In the utter fullness of an uproaring upheaval...

The maelstrom to end all messes and shambles,
The lawless free-for-all of total energetic anarchy,
Entropy crowned as King of the great hullabaloo,
That cosmic hoopla in which all hell broke loose.

Never is there to punish one for not even knowing
Why one is here in this world so much growing,
That here became all so willy-nilly going.
So, as life's rose, outspread your fragrance blowing!

Whither flowing free whether knowing, or not,
Hitherto I know not whence but am whither going,
Willy-nilly, hence that's all there is to knowing;

Hence thither forth I go on hither flowing to find
That I was ever more free to be in body and mind.

It is of Ovid's *"rude and indigested mass:*
The lifeless lump, unfashion'd, and unfram'd,
Of jarring seeds; and justly Chaos nam'd.

"No sun was lighted up, the world to view;
No moon did yet her blunted horns renew:
Nor yet was Earth suspended in the sky,
Nor pois'd, did on her own foundations lye:

"Nor seas about the shores their arms had thrown;
But earth, and air, and water, were in one.
Thus air was void of light, and earth unstable,
And water's dark abyss unnavigable."

So it is that we the living might hereby agree,
To live a being that is much more intense,
To leap toward higher orders of actuality,
To revel in the glories of this conscious life,
To attain each minute a more euphoric joy...

And to bring this radiance forth to all,
The increased intensity of free experience,
And to build on it, etc.,
Ever growing; forever, amen!

HISTOIRE D'AMOUR
LOVE STORY

Un homme tombe en amour avec ses yeux,
A man falls in love through his eyes,

Une femme travers ses oreilles;
A woman through her ears;

Plus tard, il renverse ...
Later it reverses...

Une femme prend note de tout ce qui sera fait,
A woman notes everything to be done,

Mais l'homme ne connaót pas le voir un.
But the man does not hear the seeing one.

Mais il ya encore de l'espoir ...
But there is still hope...

Comme dans le mariage
As in the marriage

De la femme aveugle
Of the blind lady

Pour l'homme sourd.
To the deaf man.

THE QUICK AND THE DEAD—
ON THE LAST AUSTRALIAN MUNDI

Graybeard (Greg) headed for
The waterless Mundi regions,
Where the winds and the sands
Sculpted and streaked the rocks,
And where the Knights Templar
Of the armor plates
Would be at a disadvantage...

There he waited and looked up
At the sharp white stars.

Soon his pursuers would arrive,
For he had let it out
That this was his destination;
However, all was never as it seemed.

On the Last Mundi, or was it Tuesday,
Greg (Graybeard's alter ego) was walking
Along the windswept plain of the Mundi
On his way back to his camp at the large rock,
Returning from a hike in the mountains.

It had been a good day with nature;
He already felt somewhat primeval.

It was almost dusk
And so the stars of home
Would soon shine above and beyond.

It was good to get away to see and learn
What more that this life was all about.

What's that!
A mad rabid dog ran out from the shadows,
Heading crazily but swiftly toward Graybeard...

How did he know all this?
Of what is a human made?

Would an acute fear response
Give him a good shot at staying alive?
Should he confront or avoid?

In this case he would have to
Try to avoid by flight
And then perhaps confront by fight,
Which is really more like a freeze,
There being flight, fight, or fright.

We are actually hardwired to flee first,
But if overtaken we must defend,
Although a trancelike passive state
Of being filled with fear is also possible.

Within seconds Graybeard was primed;
His pupils were dilated
And his respiration had sped up.

He stopped producing saliva
And sweat poured out all over;
His blood rushed away from his stomach
To soak his brain and muscles
In nutrients and oxygen,
Energizing him for what lay ahead.

He froze, watched, and listened
But only for a second.

Light waves flashing off of the dog's teeth
Had passed through his eyes
That could now see all the better.

Electric signals had entered his brain,
The visual information
Routed to the opposite sides,
This depth perception helping him
To better locate and keep track
Of the oncoming and insane assailant.

The sound waves of
The dog's growling and barking
Crashed against the tympanic membranes
Of his ears and were on into his brain
As sounds to be processed.

Yes all this as well in a second or two.

Within milliseconds,
Neurotransmitters chemically

Ferried electric signals
From one neuron to the next,
Spreading the latest news of the dog
To a quick response unit
Born of ancient times.

The sensory information had funneled
Deeper into the brain for further analysis.

Graybeard's vast network of neurons
Lit up like a Christmas tree.
The ultimate decision would be made
By the amygdala—the "fear" center.

Would there just be cause
For a temporary alertness
Or should there be
A full-fledged fear response?

The dog was going wild;
There were no trees to climb;
There was little chance for escape, but....

The amygdala sent a siren sounding
Through Graybeard's brain,
Having cued the locus ceruleus
In the brain stem to release gobs
Of the neurotransmitter nor-epinephrine.

Twin brain structures
At the bottom of the head,
The cerebellum,
Considered various attempts
For escape or of self-defense.

All of Graybeard's ancestors
Had now arisen to heed the threat.

The brain stem had sent
An all-points bulletin,
This constricting his blood vessels
And inhibiting all ordinary
Parasympathetic nervous activity.

His throat tightened
In case a scream was necessary;

His body was preparing for the worst.

This real danger was
What his formerly safe life
Had come down to.

To stare death in the face
Was now to live twice.

The dog was fifty yards away
And bearing down upon him,
Its nature having gone wilder than wild.

The spinal cord had aided the cascade
Of the acute fear response
To all the corners of Graybeard's body,
Activating the peripheral nervous system
Of his arms and legs, among other senses,
To attend to stimuli
Of the new and dangerous environment.

Greg threw a few stones;
No effect. Some more; nothing;
They could not halt
The foaming rabid beast.

The flight signal had reached his muscles,
Their fibers already contracted
To increase his running ability;
Heart and legs racing,
He ran and then looked back
To see the savage dog gaining on him...

He threw a larger stone;
It even hit the dog,
But there was no overall effect.

Graybeard reflected on
All his years on ToeQuest,
Wishing that he had said "boloney"
A few more times.

The crossroads all went nowhere;
The signposts pointed to oblivion.

The vicious dog was almost upon him,

So he stopped and waited
And planned for the fight,
Having but a second.

He wished that he
Had brought a weapon along;
There was not even a stick
Or a branch lying about.

He recalled his bevy of girlfriends,
But for the one he had given Austin,
For she was not much of a scientist.

Eternity called out Greg's name.
But this was a wrong number,
For he was now totally
Graybeard the Magnificent.

The foaming dog leapt for Graybeard;
Even one bite wold be fatal—
Graybeard's sturdy hiking boot
Caught the dog in the throat
As he kicked toward that vital area,
Stunning the dog
And sending him to the ground.

Just as the dog was about to recover,
Graybeard dropped a knee
Into his head and crushed it,
The poor creature's brains
Splattering all over the ground.

Greg's body and mind still swirled
With the rapid firings
Of the acute fear response
But he eventually calmed down.

There was a sour taste
In his mouth—
His salivary glands
Were turning back on,
A good sign that his life
Was returning to normal.

He walked back to camp
And drank the beer

That his glands had further requested.

Greg wondered how ninja Graybeard
Had accomplished the kill, then thought,
"Thank you, evolution and natural selection!
You made me what I am today."

Another dog arrived, a tame one.
Greg talked to it like a friend.

MUNDI EPILOG

'Twas the Pope's highest Cardinal [Sin] himself
Who'd ordered the assault on the Gray One,
And so the end was to be at hand,
But on the other hand Graybeard had algae,
And had flung it into the eyes of his pursuers,
Then patted the end of his horse.

The knights faltered and gave up the chase
But the spiritual chasers appeared in their stead,
Stating unscientific theories,
Such as "God is".

And then?
The Spiritual Chasers of God Arrived.

Meanwhile, science was (re)written
To say that *some ancient wise men*
Had long ago discovered
The weak and strong nuclear forces
While thinking about earth, air, fire and water,
Then correlating it to consciousness.

One of the ancient ones discovers a photon,
As well as the entire electromagnetic spectrum,
The strong force and the weak force,
Which corresponds to thought,
But they called it "Taurus".

The spiritualists wanted it all;
They had to ordain themselves as "special",
Above and beyond all the rest,
For that way they could be deserving
Of even more reward in the afterlife,

All this born out of their pride
Of their very own Divine Creation...
That they made up.

Greg had befriended some of the spirituals
But had to use logic on some of those remaining...

From the top of a large weathered rock,
Graybeard cried out, "Where is God?"

Each spiritual answered in turn:

"He is between our heartbeats and breaths."
"He is life."
"He is the universe."
"He is love."
"He is everywhere."
"All is of His illusion."

Graybeard answered,
"You have said nothing but life is life,
And that the universe is the universe and so forth,
Just equating one real thing
With another name of an invisible thing
That is even quite undefined in the first place.

Who are you all that makest all of these words
Plied upon and on top of what is?"

"We are the spiritual chasers of God;
We label Him as anything
And everything we choose."

Old Gray, looking a bit like God himself,
With his long gray and flowing beard, continued,
"Are you human mammals of such recent vintage
So extraordinarily important and special
In the whole entire scheme of things
That took so many tens of billions of years
To stumble along in such as haphazard way?"

"Yes."

"Do a trillion stars exist just to illumine your night?"

"Yes."

"Do forty million species thrive just for your delight?"

"Yes."

"And is all of space just for show, to glorify you?"

"Yes."

"Did the supernovae stardust showers
Of the atomic elements write the names
Of future humans across the sky way back when?"

"Yes."

"Does every atom exist and spin to service you?"

"Yes."

"Did Proto-men, and before them, and all, live,
Die and suffer only for your promise?"

"Yes."

"So then every dinosaur,
And more was condemned
So you could gain a space
To live, war and kill?"

"Yes."

"Does the sun shine with you in mind?"

"Yes."

"Was Heaven's Shrine built just to await your coming?"

"Yes."

"Oh my, religious ones, how vain and proud you all are!
What hubris, conceit, self-love, and vanity
Have you to claim such full self-importance
To demand so much from the universe...

"That you would even claim an angelic vapor that
Drives a living being, provides character,

Morality, and consciousness on top of
A burdensome, fragile, and expensive
Organ such as a brain ne'er to be used?

"It's a silliness born from exaggerated
Self-worth, an invisible hilarity—
Becoming a merciless indoctrination.
May you all soon recover your humility."

As such spoke the humble Graybeard to show
The truth of what we all are: mammal, organic,
Past narcissism and self-adulation
To the bio-electro-chemical organism
Evolved upon a planet near a star
In the long and winding mindless way of
Slow time, dust, and selection by death that
Sifts the best from the rest: evolution.

And such did Analog once observe that
The creature out thinks the creator,
With inferior tools, to imagine a
Much more peaceful and enjoyful world,
And that it is emotion that creates
Delusions of heavenly scenarios of
Creation, and an existence beyond death.

These are lessons in humility to all
Mammals grown so high and haughty...

So... enjoy it all as though you will never
Know it again; for who is to say that you shall?

Only one spiritual was left at the base of the rock,
But she had Graybeard surrounded...
With her words on evolution symbolized by the Bible.

"That's it," said Gray Newt, "I'm gone",
And so he fired up his jet pack
And launched himself into the sky
Toward the white patches of vapor and fluff.

Some of the aboriginals now thought him a God—
And so he became the legend:
That God was an old white-haired
[Gray bearded]
Guy sitting on a cloud.

MUNDI MEMORIES

Greg walked to the mountain and back,
Sitting safe on some lone rock for the lack
Of any other seat to pick but that of his own;
Wherewhence he slept, thinking he was all alone.

As there he lay asleep so peacefully in repose,
Some dogs wandered by and licked his nose.
And while he turned untossed, a kangaroo
Of boundless flight, hopped over him, too.

The Great Equalizer stalks all creatures made,
Lying ever just 'round the corner in the shade,
Taking both human and the beetle as one,
After their lives are spent from rolling some dung.

THE UNDESIGNED

UnTruths are but great amusements and fun,
As we are the undoing of the Perfect One:

Some of the Angels supposedly turned 'bad',
From the perfect planning of creation they never had.

Adam and Eve Sapiens goofed in no time,
As of Intelligent Design there wasn't a sign;
Noah's progeny screwed up right and left,
Since they were of the Master's hand bereft.

Of the Ten Commandments few were impressed,
As no such thing came down from the crest.
Two thousand years of folly now from redemption
Were no picnic: 'Design' was from evolution!

Evolution, driven by natural selection,
Is a design without a designer.

TEN DAYS TO WHITEHORSE
A Yukon Quest First;
First woman to cross the finish line:
Lorrina Mitchell, 1984,
aka LabelWench

A thousand miles through the northland white,
With its four peaks and endless trails of fright
I'll pass, with my dogs, four hours off and on,
Over frozen rivers and the tundra far beyond.

The moon will light the night like the day,
As the winds whip my thermometer away—
Minus forty and lowered by the wind chill,
As I descend the mountains and the hills.

The Yukon is a quest to know what we are
And were, as when we rushed for gold afar,
Flying past the furthest bar, over-shone by stars.
What strength will I find where the eternal are?

What icy river might sweep me into cold embrace
Before I can finish this animal and human race?
What exhilaration may drain to solitary disgrace
In the middle of some unchecked, pointless place?

What the whether of the extremes I can weather?
Whither whence I slide heretofore wench-untethered?
What fate awaits in this land of legend and lore,
Where no woman sledder has ever gone before?

I'm away and off from Fairbanks forevermore.

EMOTIONAL DECISIONS

Emotional decisions can make one happy,
In everyday life as lived,
Even for such as what the All should be,
But there they can block
The way of the light of truth.

THIS TOTTERING EXISTENCE

So called "empty" space is vital,
For that's where there's the recital
That forms and plays the tunes of reality,
This grand cosmic symphony,
As existence fluctuates with the non,
Those causeless waverings of undulation.

It was once thought that the shove
Of this total energy was of
The order of 10**120 orders of
Magnitude above.

Well if that were so near
Then we couldn't even be here;

It was the worst calculation
In all of scientification;
So we weighed the universe,
Summing all of its constituent verses.

The universe weighs nothing at all!

This too since we found that
Our universal space was B flat—
Not just via the 60 degree angles
Of a very small triangle,
And not by even using stars,
Nor one that went from here to Mars
To Venus and back,
But all the way back
To a degree of the CMBR,
Which represented 100,000 light years,
For which we measured the curvature:
The rays didn't converge or diverge.

The ultimate of this geometry
Is that being flat is the beautiful symmetry
That leads to yet another beauty: zero.
The ever returning, conquering hero.

ON EARTH

The sun fills the waking and breathing world
With the fire of her imagination.
In poetry the sun is the power behind the mind;
The moon, planets, and stars are symbols too.

Sometimes intellectual beauty is bright
And ideas gush from the eternal flame;
Sometimes it fails when the shadows of clouds
Dim the clarity of thought now and then.

Quenchless, boundless, ever bright and burning,
The mind's light searches every dark cavern,
Probing, imagining—its beam alighting
Upon the earth or high atop cloud mist,

And melts, with heat, energy, and desire,
The fog of lone reason and pure passion,
Burning it away, soft dissolving it
With the love of life, earth, mankind, and star—

From which comes adventure, friendship, delight,
Joy, success, triumph, and lasting gladness
Throughout the sun's journey into the night,
When stars shine on mind—suns they also are!

The moon fills the sleeping and breathing world
With the icy coolness of chaste reason
Unaffected by deep burning passions
Although sunlit to glow in its wan light.

Reason, unsteady as the variant moon,
Oft does not rise in the night to guide us,
And deserts us in darkest times of woe;
We are alone on a black cloud-bound night!

Else the moon hides in the bright light of day,
Or is lost behind an overcast sky;
But moonless nights take us beyond reason
When the stars excite us with their lights.

Yes, inspiration returns with the stars—
A thousand ideas beckon from afar;
Ideas wink like fireflies on the mind's meadow—
As starlight they stab the darkness of nought,

Until star-like Venus rises near dawn.
Goddess of romantic love and passion,
She captures us within emotion's swell,
While comets flash and confuse the wild sky.

Soon intellectual beauty returns,
Borne on birds' wings as song into the dawn,
For all human music is but a part
Of earth's ancient melody and rhythm.

Imagination now soars past a day,
And into the season of spring's fast growth;
The shade is deep and cool, like the ghost of
Winter passing—gone but still remembered.

WHEN YOU ARE YOU NO MORE

Who is concerned about an nonexistent brew,
A nonconscious state when there is no 'you'?

No one, for none spent the billions of years
Before they were born in a state of tears
Of anxiety and apprehension of not being here.

QUANTUM COMPUTATIONAL UNIVERSE(S)

By traversing all possible initial arrangements,
The Everything/Possibility/Potential
Came upon our particular solution,
Perhaps among many other good ones.

Whether all universes exist in the actual
Rather than just some promising ones
Becoming from the possible
Remains to be worked out.

Did mammal consciousness
Bring ours into being somehow,
This being as the universe knowing itself?

OLD AUTUMN

The glow-worms, fairy stars come down to ground,
Gleam the shadowy woods through summer's round;
Then fall's leaves flutter through the quiet air,
The autumn being the sunset of the year.

The rustling of the trees comes to my ear,
In this,the most mellow time of year.
The harvest brings fulfillment, yearning too,
For autumn is both a smile and a tear.

Each year in October Jack-in-the-Green
Has a chilled rendezvous with Old Autumn,
Who colors the leaves that Jack made verdant
A season ago. They meet out in the woods,
Although never in the same place, for seasons
Come and go and meet each other as they may.

This year Old Autumn was a little late,
So Jack-in-the-Green sat down on a stump.
Jack pondered his disappearing green youth,
For someday he would have to take Autumn's place
And perform all of his withering tasks.

A few days later Old Autumn came by;
He gave unto Jack a cheery greeting
And a warm embrace that marked summer's end.

He gazed fondly at Jack, his younger self,
And saw the vitality that was once his,
And said, "Once I was young; once I was you!"

"I know," said Jack, *"Do you remember how
I refused to believe you, saying 'no'?"*

"Yes," remembered Old Autumn, "very well,
Like the time I met the Old Man, Winter
On a snowy December day long ago.
He told me that he was my older self—
But I didn't believe him! Told him off!

"True, I was already feeling my age
But after seeing the old white-haired geezer
I felt young again! Yes, he knew me well."

"Right," said Jack, *"so I made a little poem:*
"When younger, I knew not my elder same,
But when older I told my younger same
That youth must be young; he knew not my name!
It was my younger self who was to blame!"

Swallows twittered in the skies as sprightly
Jack-in-the-Green picked a ripening gourd
And gave it to Old Autumn, who encouraged,
"You won't have to meet the Old Man until
You take my place, for only I can see him—
After I take down the last of the oak leaves.

"For now, the Old Man sends but his errand boy,
Jack Frost, your twin brother. Hi ho, here he comes!
Aye, young Jack, this is the rarest of days,
For the three of us can be together
But once a year on this bright day / cool night."

"The Old Man is so lonely, is he not?"
Asked Jack-in-the-Green, *"for he sees only you."*

"Yes. Old Man Winter lives cold and alone;
He never sees the fair maidens of spring
Who reinvent the natural world each year."

There is a chill in the air as Jack Frost arrives
And sings out a greeting: *"Hello my brother!*
Hello Old Autumn! It's going to be cold—
Our first frost, but don't worry too much—
It won't harm the pumpkins any at all."

Old Autumn sighed and quick replied: "Good.
Now the rest of the leaves will crack and fall
All the more due to the ice in their veins;
Yes, they'll fall like the illusions of youth,
'Lying carelessly on the granary floor' and
'On a half-reaped furrow sound asleep,
Drowsed with the fume of poppies', as Keats wrote."

Composing himself, Old Autumn continued:
"And for those of you who think that 'warm days
Will never cease', let us ever remember
Dear Johnny Keats, who died so young, at 25;
However, he lived and saw more than some
Of us might hope to do in a lifetime."

A shiver ran through Jack-in-the-Green,
Hence he said: *"It's cold; I must go now, for*
Summer passed away in his sleep last night;
Autumn, sweet and plump, carries his offspring.
The year dies in the night; ghostly winter looms;
Lo; the flower is already in the seed."

"Well done, young Jack-in-the-Green; quick, go,
For soon enough comes your autumn of care
Sobering into age, thence into
The pale white winter of death,
Though not yet your warm indolent summer
Of contentment lazing into middle-age,
But surely past is our crisp,
Flowering youth-spring of joy!

"Such then, comes the end of summer's dreams,
The blanching of the grassy banks of streams,
But all fragrances my elves remember
Through their long sleep in the winter embers.

"The blossoms fall, showers of fragrant beauty,
As leaves fade, while the bulbs store up energy.
Nature's floral dreams grant this destiny,
For these leavings enrich earth's potpourri.

"Flowers lay their heads to sleep in soft beds,
Blanketed by webs of gossamer threads;
My elfin creatures cast their spectral glow,
As winter stars—floral twins—start to grow.

"Later, when surely all the world is dead,
An elf will stand atop Old Winter's grave
And say, ''tis not dead', and by magic bred
Make Snowdrops flower in the tomb's heat wave."

Once I, the author, ventured outside at
Four on a dark frosty October morning.
It was so quiet that I could sense the
Cosmos as it played rhythm to my beating heart.

I saw a preview of the winter stars:
Orion, you are so high in the sky—
There for only the astronomer's eye,
As all those meteors go flying by.

Then I heard a rustling sound in the leaves
Around me—a skunk perhaps—but no,
It was the sound of many falling leaves.
I knew that it must be him, Old Autumn.

He was out there somewhere. Then I sensed him
Going by, for some of the leaves on the
Tree right in front of me broke loose and
Floated away, hitting some other leaves
On the way down, making that rustling sound.

Soon it started up on the next tree, and
Then the next—and so I could very well
Follow the path of Old Autumn making
His rounds in the misty October morn.

Chrysanthemums drank the mellow day,
Falling petals carried the light away.

The weed-flowers grew, marking autumn's track,
The blossoms that almost brought the spring back,
But winter's white death wrap was drawn over,
Smothering the earth's last warm sweet odour.

The autumn fog enswirled, the mist upcurled;
Into nothingness the wisp slow unfurled.
November flew by, a colorless dearth,
And December, amid death, a festive birth.

Youth and Beauty made agèd Winter mourn
For Summer's grain—the waving wheat and corn,
For Old Autumn, withered, wan, had passed on,
Leaving the earth a widow, weather worn.

Long since have the winds scattered the leaves
Of the trees to make of them a
Burial shroud for the flowers that died
Grieving at summer's passing. All is death.

The fall is now nearly lost to memory.
Winter is summer's ungrateful heir,
Squandering his riches and abusing his gifts.
It's not Old Man Winter's fault, but his duty.

Summer lies underground now, forgotten,
Silent and crusty, covered by winter's

Stern mantle. Only April's tears can make
His grave green again, in the spring-tide.

As seasons pass, the world comes to our door:
Spring sings through the wingèd troubadour;
Summer calls with the rose, 'midst the wood-lore;
Autumn crows, plump and sweet, through frosty hoar.

Joy and exuberance are spring's largesse.
Sunlight, warmth, and growth are summer's bequest.
Autumn brings wealth with the mellow harvest.
Winter's fruit is peace—its bounty is rest.

Past us is the flower of spring's soft breath;
Though not ended our summer of promise;
Soon enough will come the autumn of care;
Beheld, at last, the dull white shroud of death!

March, April! spring! We'll reign as we May there,
Between June and her sister, September,
Then prolong the fall, till November come
December, when we can sweet Remember.

In the whisperings of the after-years
The winds of time slowly dry the tears;
Nor would I take back a single drop, for
From those tears the flowers grew without fears.

In spring we rise from the garden at birth.
Summer blooms long with the roses' fresh mirth.
Autumn creeps in—we wither on the vine.
Last comes winter, when we return to earth.

———————

Drive my dead thoughts over the universe,
Like wither'd leaves, to quicken a new birth;

And, by the incantation of this verse,
Scatter, as from an unextinguish'd hearth
Ashes and sparks, my words among mankind!

Be through my lips to unawaken'd earth
The trumpet of a prophecy! O Wind,
If Winter comes, can Spring be far behind?

Percy Bysshe Shelley
English poet (1792 - 1822)

RECOLLECTIONS OF WAR

A fading eagle flew frozen in fear
Past deserted flowers in desperate land,
And a rising earth halted for a hasty madness
As time awaited a dead sun.

Remember now the beginning, one fine day,
When we came out of nowhere!
In no cradle birth, one thunderous heartbeat
Separated animal from plant,

And we stood up straight one day,
Our minds still a drunk's uneven crawling;
Later, in the breath of life, we knew that
A churchyard must yawn now and then.

But now we are helpless—
We must fight to our worthless deaths, dying,
Screaming forgiveness, but die as we must
When peace is a barren land.

Daily now one grips less firmly his last integrity,
The essential life slips.

Where are the grown men, stuffed and rigid?
Where are they? Where?
They are so silent and meaningless to us now.
They are no longer with us.

And throughout the aftermath
We could almost grasp it in our dreams,
And hope that we might live to die
Far from the River of Perfumes.

Meanwhile, we are dying to live everyday
As we surrender our souls.

Around us we see the bodies—
They lie upon us; they died among us.
Rising to our last stand we look:
Where are the grown men, the old men?
We thought that we were loved then,
But we've been betrayed, sold, lost ...

Shall I try for fading woods,
Scrambling over the trails, searching for my life?
I'll flee and fly over the leaves of yesterday—
They crumble before my eyes.

And there I'll come out of it all
With firm desire to laugh, love, and live—
There in a hilly grove
Near swelling stream by daisies, grass, and tree.

Once more I escape the horrid death
As the grown men approach;
I try to see my way past
The swiftly moving figures of the human race.

Even now those men with guns so loud
Are silently dying in the strife.
Living in a time nigh for sighing,
We rise for dying.
Can this be life?

Of course all this it was our duty to bear;
We bled our blood; we served.
And during the lull of the monsoon rains,
I begin to drink, to honor my life—
To hope, as dawn comes,
Much like a Chinese painting—
Too real to be true.

I wake the artillery-man,
And cross the Song Ba to disarm the claymores.

Now it is lovely April and we're dying
On this fine day in the time of our life.
Slightly sighing for crying Charlie,
My bayonet blazes in scarlet, in death,
And yet another hasty man
Gropes for the earth and escapes this horrid life.

But there, on a cloud of thought
We fly by their ways with a life for ourselves.
And then they wither with the wind,
Those thoughts that once echoed,
Where they once were teeming, fighting...
The forums now emptied.

There is only room to say
"Let us kill him," as wrath's way becomes us,
And there in the cells of a brain
Where currents of feeling once surged,
The mind's will falters, and waivers
Between the Emotion and the Intellect.

A shrill siren chilled with ill will,
Then when he was yet young and fine—
Houses were crumbling, streets were heaving,
People were weeping, dying;
And others wished to live,
From brothers to mothers—all lived but the father;
Can you see the tears in the young one's eyes
As the deathman cometh?

The love and the feeling were nowhere,
The men motionless and rigid,
And too the air was not worth breathing,
But was filled and smothering,
Leaving the men breathless, helpless,
And of course so lifeless.

The blight was so deathtaking;
The sight of goodness never so breathtaking.
Once in awhile I'll wince in a smile for truth,
Cringe at the fringe of love;
It is my dream,
A star shining somewhere in the universe—
I can see it there in all of its dimness,
Through the plight of my brightness.

It is there forever and still;
It is there while the thinkers thought for ages,
As dreamers dreamt time after time,
When hoped even the hopeless,
As slept the sleepers into oblivion,
While philosophers pondered infinitum,
As wept the weepers for a long time,
When pitied even the pitiful . . .

All that I saw on Earth was lost.
There hated the loveless in the wasteland.
There the dying lived for a lifetime
As all the wise men greyed and died.
So now I'll let my 'enemies' grow old

As my wine yet flows sweet and pure.

Here comes the slush of doom
Seeping over us,
Belching with contagion.
The pleas of the corrupt fly out;
They cry out; their lives are snuffed out!
The Good Friday mourners yet weep for man,
For everyone, for eternity.

At life's end
The silent men array themselves,
Finally—
There for the asking
In the stead of the dead,
Prisoners of themselves.

Cautious Pilate ponders,
As there my star shines in the springtime of life.
The star is a beacon in the night of terror,
Fading in the search for the valiant.

How can I live, how can I die?
Look around—there are other worlds!
See, the grass is high and green
On the far side of never.

Find for me the sun shining,
The streams flowing,
The forests, the fertile meadows.

The soldiers moved slowly now
To make their lives last,
A searching band—
And fighting has flared on the border;
Now hurry death or hurry darkness.

Deciding at last, I made an easy day of it,
Staring life in the face, indulging
In a vast wonderland and wilderness
Of childish fancy and fantasy,
And I laughed a lot louder then,
Feeling no need to weep in pity for them,
Or to cry for the scoundrels
Who would grasp at life from graves in war.

It was then that I saw the life,
The awe, the infinite,
The good, and my end.

To see where my youth
And laughter could go,
I lived and died to be free;
My mind took no mind;
Yes it was good to be loved then,
To be young again.

The TOE
*The conundrum of an enigma
wrapped in a riddle revealed*

We're of the endless forms most beautiful,
Stunned that our glasses to the brim are full,
Life's wine coursing through us, as 'magical',
On this lovely, rolling sphere so bountiful.

What IS never began; it's the causeless,
For you can't get anything from Nothing;
Yet there's a zero-sum balance of things,
And nothing to make the things of, no less.

If there is a way to make things from naught,
Then this is <u>something</u>: capability;
So we're back to an IS eterne, ne'er wrought,
Having no choice, no option; must be.

No matter if it's a wave or a field,
A substance, or Possibility's yield.
As undirected, all eventually
Forms from it, 'beauteous' or 'beastly'.

Since outputs always have inputs, so true,
Then what, we wonder, should we try to do?
It's the other way around, of brain stew,
For it, time, and the universe does you!

THE LOVE LIFE OF THE GLOW-WORM

The day pours life into roots with sunlight.
Flowers bloom, showering us with delight.
In a blossom, a glow-worm blinks its light,
Kindling the flames of a romantic night.

The glow-worms, fairy stars come down to ground,
Gleam the shadowy woods through summer's round.
During the gloaming they warm up their lights,
And then metamorphose wings for their flights.

The dusk deepens, night's pot of tea steepens;
Silence descends, as when a gift opens.
Eventide rises. On high, Scorpius camps;
The eyes catch stars, like fireflies in lamps.

The sky is lit, a twinkling promenade,
Of mating calls from luminated pods,
Tracers pulsing wild, searching thoughts that smile—
From fireflies named Winkin', Blinkin', and Nod.

Flashing desire, the glow-fly twinkles across
The starry summer sky, love's energy unspent,
Searching through the darkness, with passion's might,
For the beacon of her consent—the surging sight
Of love's pulsing, green and yellow light.

The reply: "Yes, oh yes", alight, she winks, to woo.
Now he becomes a firefly, as at once she does too.

"Come light your lantern and mine with good cheer;
We're magic lamps—our spirits dance in there.
Our beginnings and ends are of nowhere,
So let's radiate, since for now we're here!"

In a closing flower, they together make their stead,
Blinking, winking, in the seclusion of its petal bed.
This dance of light and love—their honeymoon,
Brightens the night—till it looks much like noon.

Their jolts and bolts, surging, merge in currents,
Sweeping back and forth, as they signal delight—
Fires luming and oft reluming
The flames of love, in electric hugs—
For they have by now become lightning bugs.

A LOVE STORY OF THE EARTH AND THE MOON

As the moon, challenge night and gain the light;
As the rose, suffer the thorn—gain the fragrance;
Of life, surrender to live forever—
Enlightened more than a thousand suns.

I am thy moon, thy constant satellite,
Thy crystal paramour of day and night.
Above and below, and within thy sight,
I whirl around you in loving delight.

In a magnetic dance I whirl and twirl,
Attracted to you, oh liveliest world;
Around you as a necklace I'm aswirl—
Wear me as thy crystalline gem impearled.

Wherever thou orbits 'round Apollo,
I must twirl and whirl, hurry and follow;
Dust I gather, meteors I swallow,
Ranging far and wide through space not hollow.

Thy romantic beam, as Cupid's arrow,
Pierces my heart and kills my sorrow,
Injecting life and love for tomorrow;
Henceforth I'll shine with this life I borrow.

Around you I whirl, a necklace of pearl,
Trailing afterimages of my world,
Adorning you, thy bosom bountiful,
With crystalline gems of another world.

Oh moon, thy Earth would wobble like a top
With your steadying influence not,
In turns quick of searing and freezing ruins,
Unto dying soon, without you, oh moon!

As twin planets, our orbits must convolve;
Into each our tidal motions dissolve.
Around a common center we revolve—
The focus from which our passions evolve.

As twin planets, each other's way we pave,
With the push-pulse of the graviton wave.
We're captured, but not as each other's slave,
For to the sun our orbits are concave.

To your lines of flux my path I align—
I'm your constant paramour, crystalline;
Your world pours life on mine, on mine!
Dearest Earth, I must be thine, must be thine!

A magnetic beam emanates from thee,
Attracting me, holding me, kissing me;
Tidal love washes freely over me,
Linking you and me for eternity.

Basking warmly in your reflected light
I'm bright, oh so radiant in your sight!
In the love and light of your spirit bright
I need not ever face the endless night.

Your vibrations travel without a sound,
Circling from all directions to surround;
This affection touches me 'round and 'round
And closely binds me to you—I'm love-bound!

We're as different as midnight and noon,
Yet drawn close by the force of Earth and moon;
As lovers we merge in a sweet eclipse,
When world meets world as a kiss on our lips.

Oh as your shadow of love covers me
I am full, so full in the shade of thee;
When we overlap, that union is us;
The you is in me, the me is in thee!

As moon and Earth we bathe in radiance,
Cleansing our hearts in love's grand alliance;
Around and around each other we dance,
Entranced by the whirl of our dalliance.

My blood runs warm with the sun's heat at noon.
My spirit is swept by thee, swelling moon.
Space surrounds us. The tides flow through us.
Global rhythms are always playing our tune.

NEW 9TH PLANET FOUND!

Poor Pluto's been banished to the underworld,
Charon rowing him to the land of the forgotten.
Schoolchildren petitioned for his return,
But he was voted off of the solar island.

Memory's crutch for the order of the planets,
Is now just "MVEMJSUN"—
Old Pluto tried so darn hard, its position
Now even closer to the sun than Neptune's.

Well, many have searched for quite a while for
The next planet without any success—
There have been hoaxes, theories, and some ghosts;
Yet, I have firm proof of another planet.

But, first, a review of some poor attempts:
'Vulcan' was spotted very close to the sun,
And 'existed' for about five days,
But now is relegated to the Star Trek World.

Another 'Vulcan', impossible to see,
Being 180 degrees away from Earth,
Behind the sun, was seen in the movie
'Journey to the Far Side of the Sun'.

Could an asteroid like Eris be a planet?
Nope, 'tis not allowed, although all of the
Debris between Mars and Jupiter
Could have come from an unstable planet.

Nice try, but it's not out there anymore,
And any planets of other solar systems
Don't count, nor does Planet Hollywood
Or Daily Planet or any other restaurants.

Perhaps there's another planet way out,
Beyond; that may be so, but, no matter,
Though it may become the 10th planet, since
I have found the newest 9th, with no doubt.

The 9th planet does follow an orbit
Close to Earth's, ever falling toward the sun—
It is right under our nose: It's the moon!
But, wait, you say, it is Earth's satellite.

Our moon is unique in the solar system—
It's not captured by the Earth, but by the sun,
It's orbit being everywhere concave to Sol.
(Thanks to Isaac Asimov for proving this.)

Never does our moon fall away from the sun,
For it's attracted to it about twice as much
As it is to the Earth, although the moon and
The Earth do form a double planet system

That revolves about a common point that
Happens to be inside of the Earth.

THE 4TH YOGIC BODY

You just have to love this guy!

Yogi Berra, the great New York Yankees catcher,
Said many sayings that seemed to make sense
But really didn't, or maybe they did,
Like about a restaurant:
"No one goes there anymore; it's too crowded.",
And about a ball field:
"It gets dark early out here", plus
"If people don't want to come out to the ball park,
Nobody's going to stop them",
Or recurrences:
"It's deja vu all over again",
And himself:
"I didn't really say all of the things I said",
And many more great unsayings
Just as meaningful or not
As some that we see online.

In his 3rd Yogic reincarnation,
He was a coach and a manager,
And is now an elder statesman in his 4th.

Yogi Berra was simply nicknamed "Yogi"
Because he looked like one.
Nor did he disappoint in this area
But came through time and time again
With his enigmatic observations, and
This was his overall Yogic self.

INTO THE LANDS OF THE GODS

Towards the Gods Far and Unknown

My reverie took flight, with autumn's sight,
For I was abstracted, entranced, and light.

I beamed to the site suffused with insight—
The solutions are deep within the mind,
Reachable by dreams of the lucid kind.

I flew south from my home, in New Hamburg,
Over the Hudson river, toward Newburgh,
Past Chelsea, and the great Storm King Mountain—
On philosophical aspiration.

A wake of leaves trailed behind, like a stream,
While I gathered clues, through my musing means.
My design, in this vaporous pipe dream,
Was to converse with all the Gods who seemed.

If Fishkill's and Peekskill's murderous names
Had not been token enough, there soon came
A sequence of locales that seemed to be
Ominous in their triple proximity.

First was Sleepy Hollow, the haunted land
Of the gambols of the headless horseman,
Then the Gate of Heaven Cemetery,
And the surprising Town of Valhalla—
A bright afterlife of an old-time place
Of shops built right up against the road race.

I stopped to rest, well away from the maze,
Dazzled by the lustrous autumnal haze,
In a warm day's musk, before twilight dusk,
Near shining gates, toward the unearthly sod
Of the refulgent Graveyard of the Gods.

Over the stream, there was an arched bridge thrown.
Then I knew I'd gone beyond the known:
For in that span, each piece was a keystone.

I questioned two luminous angel goths,
"Where be the mythic Graveyard of the Gods?"

They looked askance, then smiled and pointed past,
"It's just beyond the land of epitaphs."

Remembrances

THE CEMETERY WAS WHERE THE DUCKS WERE FED,
WHERE TWO FRIENDS FEASTED ON WINE, VERSE, AND BREAD,
AMIDST THE FLOWERED TREES AND QUIET STREAMS—
THE HOME FOR BOTH THE LIVING AND THE DEAD.

WE LIVED AT ONCE, AWARE THAT LIFE WAS DEAR,
OFT SMILING AT HEAVEN AND HELL WITHOUT FEAR;
YES, WE HAD SOME LAUGHS, GAVE TRUE LOVE, AND MADE
LIFE BETTER—FOR IT WAS NOW AND WE WERE HERE.

HERE THE GRAVE-SIGN OF THE FOUR ELEMENTS:
FROM THE FIRES OF STARS TO THOSE OF THE CREMATION,
HE HAS BREATHED, FLOURISHED, AND DISSOLVED:
LIFE IS ASHES TO ASHES, STARDUST TO STARDUST.

OF AIRY WINDS, VAPORS, AND A SOFT EARTH,
HE RESTS, AT LAST, UNDER THE SPINNING SKIES,
THOSE OF EARTH'S SUNNY DAYS AND STARRY NIGHTS.

THE SYMPHONY OF LIFE PLAYS FOR THE DEAD:
ALL THAT WE KNOW, EVEN THE LOVELIEST OF THE BEST,
DECOMPOSES INTO THE DUST OF EARTH COMPRESSED.
THE SONGS ONCE COMPOSED NOW LIE IN REPOSE;
OF THIS DUST THE FUTURE REARRANGES TO RECOMPOSE.

EN-GRAVED IS 'THE END' OF YOUR EARTHLY SIGH:
SIX SIDES 'ROUND YOU: FIVE ARE DIRT, ONE IS SKY.
SHOV'LING, DEATH TALKS TO YOU AT LAST, AND SAYS:
"WHAT WERE YOU DOING DURING ALL OF NIGH?"

FROM HEAVEN'S STARS CAME OUR DUST ETERNE;
TIME'S SEAS NURTURED THEE AND THINE IN TURN.
FROM TIME, DEATH, AND DUST WE THUS BECAME,
AND BY THIS, THUS, AND THAT WE MUST RETURN.

WHAT WOULD BE THE PRICE OF A MOMENT'S BREATH
PURCHASED FROM DEATH'S HAND AT THE FINAL HOUR?
ALL THE WORLD'S WEALTH CAN'T EXTEND THE POWER
THAT DRAINS THE CUP AND WITHERS THE FLOWER.

THE LIGHT OF HEAV'N DID THE EARTH ILLUMINE,
WHEN HE SHAPED HUMAN NATURE'S ACUMEN.
TEMPTATIONS HE THEN PLACED EVERYWHERE,
BUT HE'LL PUNISH US FOR BEING HUMAN!

THE WINGS OF TIME ARE CHECKERED BLACK AND WHITE,
AS FLUTTERING 'ROUND THE DAY FLIES THE NIGHT.
LIKE CHESS PIECES, WE GAMELY PLAY FOR LIFE,
UNTIL INTO THE BOX WE RETURN, QUITE!

NOW MY CUP WAS NEARLY EMPTY AND DONE;
THERE WAS LEFT BUT ONE LAST DROP FOR THE SUN
TO DRINK, OR WITH WHICH TO MAKE RIVERS RUN:
ITS FLAVOR BURST IN JOY—MY LIFE WAS WON!

NOT ALL POEMS ARE PLEASANT—SOME SPEAK OF DEATH,
OF LIFE'S END, SEPARATE BY JUST A BREATH.
I SAW TOMBSTONES OVERGROWN, UNDER SWEPT,
NAMES UNKNOWN—AND TO ALL THE MESSAGE SAITH:

READ ME, IT SAID, ENGRAVED BEYOND THE BRINK,
YOU, WHO LIVE, UP ABOVE: OF LIFE GO DRINK;
AND YOU, UNDERNEATH, NOW LYING SO DEAD:
REST IN PEACE, RELAX—IT'S LATER THAN YOU THINK!

REFRESHED, I WANDERED AMONG THE TOMBSTONES,
UNDER WHICH RESTED LITTLE MORE THAN BONES,
WHERE FROM THE LIFE HAD FLED WHEN DREAMS WERE DEAD,
WHICH UNDER ME BECAME LIFE'S STEPPING STONES.

I'LL PLAY THE GAME AND ROLL THE EARTHLY DIES,
AND THROUGH THIS WORLDLY LIFE ENJOY THE PRIZE;
IF EARTH IS HELL FOR LOVE'S ADVENTURERS,
THEN I WISH NO MORE FOR GOD'S PARADISE.

GOOD AND EVIL WERE WROUGHT FROM WRONG AND RIGHT,
WHEN, OF NOUGHT, TWIN GENII SPLIT DAY AND NIGHT.
SOME MAY THINK THAT BLACK'S MIGHT CAN VANQUISH WHITE,
BUT NIGHT CAN'T EVEN QUENCH THE SMALLEST LIGHT!

EVERY-THING, EVERY ORDER HAPPENS FOR A REASON.
YES, FOR THE MOST PART, FOR MOST SEASONS,
BUT NOT FOR THE BOTTOMMOST CAUSE THE FIRST,
FOR THERE WAS NOTHING BEFORE IT TO DIRECT IT FORTH.

YOUTH AND BEAUTY MADE AGÈD WINTER MOURN,
FOR SUMMER'S GRAIN—THE WAVING WHEAT AND CORN;
FOR OLD AUTUMN, WITHERED, WAN, HAD PASSED ON,
LEAVING THE EARTH A WIDOW, WEATHER WORN.

AT FIRST, YOU SLEEP IN YOUR DEAR MOTHER'S WOMB;
AT LAST, YOU SLEEP IN EARTH'S COLD SILENT TOMB.
IN BETWEEN, LIFE WHISPERS A DREAM THAT SAYS
WAKE, LIVE, FOR THE ROSE WITHERS ALL TOO SOON!

WASTE NOT THE TIME OF YOUR LIFE IN GLOOM'S DOOM!
BY THESE VERSES, YOUR LAMP OF LIFE RELUME:
YOUR LIVE BODY, FULL OF WARMTH AND BLOOM,
IS WORTH TEN THOUSAND LYING IN THE TOMB.

ART AND POETRY ENRICH HUMAN EXPERIENCE,
BUT THEY'RE NO SUBSTITUTES FOR THE LIVING OF IT.
LIKE KEATS' FIGURES ON THE URN, SHOULD WE LIVE LIFE LESS?
NO, BECAUSE WHAT IS DEATHLESS IS ALSO LIFELESS!

Figmentations

Into supernatural figmentations,
I strode, with brilliant imagination,
To examine all the supposed Gods there—
Some no more and some ruling everywhere.

Notions of 'God' are of the wide purview
Of the inquiring mind confined—its 'why',
That wide expanse of fables, faith, hoaxes,
Lies, imaginations, fictions, guesses,

Foggy notions, concoctions, phantasms,
Fantasies, falsehoods, conceptions,
Decrees, fiats, misrepresentations,
Dead ideas, magic, proclamations,

Wild tales, anecdotes, revelations,
Untruths, revelations, hearsay, scrap heaps,
Yarns, and fish stories, stated as beliefs
In that unseeable supernatural station,
Through faith's without knowledge ration;
These are all figmentations of the imagination.

Strewn about this great panoramic realm
Of the One possibly conceivable at the helm

Were all of the unknowable fabrications
Often dreamt up, via exaggerations,
By the human race of mammal sapiens.

The realm of such pronouncements has come to be
Superposed at the furthest edge of Reality,
Poised by the scope of some wishful thinking,
By all those dreaming and wild supposing,

Who wish for such legends to be ever
Actualized and realized; however,
These unknowns have never ever made it
Into our observable realistic habitat in any way,

They but remaining in the minds, joint,
Of the God-beholders—
Even as wildly varying viewpoints.

The Graveyard of the Gods

Without so much much as a word to say,
I passed those to whom most no longer pray,
Nor believe in, but once did, namely,
Those of the tombstones now deemed unholy:
Astrology—the God of the stars that plod,

Eternally blazed and marbled in the sod,

Monuments of Diana the Moon God,
Druid Gods, Apollo, Baal, Zeus, Wotan,
Aphrodite, Mithras, Isis, Amon,
Poseidon, Thor, and on and on, anon—
Posed in the burial ground of the Gods.

I ever hurried past the ledgering
Of those older Mythologies preceding
The formation of the Old Testament story—
Those ancient superstitions whose very
And various olden amalgamations
Brought forth to form it whole for our salvation.

I paused at that Old Testament maligned,
To mark the old but lingering lines
Of the 'knowing' of more invisibles—
The beliefs in imagined Angelics:

There were angels standing, frozen in stone,
Over the timeworn memorials' poems,
As well as atop the crumbling gateposts,
Cast as undying and near-living ghosts

Of the representations of the three spheres
Of the Heavenly host: the demigod-near
Seraphim, Cherubim, Ophanim,
Thrones, Principalities, Dominions,

Powers, Archangels, Angels, and, those final,
And the most useful—the Guardian Angels,
Who are said to protect children from harm.

There, Amaranth, its dead red leaves never
Fading on this Earth, unto forever,
Gave some color 'round the graveyard pallor
And to the dateless headstones' gray squalor.

There was a garish maroon view, on high,
Of streaking lights of an electromagnetic sky,
Heretofore never imagined by my self.
I strolled on, and into the vale itself.

The Intelligent Designer

I approached a semitransparent,
Theistic Embellishment, quite well lit,
Who was holding out an eyeball—a shove
Of His hand for me to take note of.

"Who might you be?" He mimed,
"For I am the God of Intelligent Design,
The One who was made by the signs discerned,
When the creationists noted them all, unlearned."

I answered, "I am Austin, Earth's flower,
Although not 'Powers', but 'Higher Powers'."

"Ha. Lo, they saw inexplicable complexity in Nature,
And thus they leapt and promulgated that Nature
Must have a Grand Designer of its mechanical dance,
For how could life have come about by 'chance'?"

I replied, "You're right about 'chance's' stance,
But wrong about 'chance' too, for little greatness,
If any at all, comes about by mere 'chance',

"Especially as some giant leap in one bound,
Up the sheer cliff-side of Mt. Improbable—
To find on its top a great complexity
Of something like the eye that You show me;

"However, it is actually an error to suppose
That 'Chance' is the scientific alternative
To Intelligent Design, for that's quite negative.

"Natural Selection is the means of the design,
For it, unlike a one-shot 'chance', being not in kind,
Is a cumulative effect that ever winds,
And slowly and so gently climbs

Around the mountain's other side, behind the sight,
To eventually arrive at the great height
Of complexity—from which we can then view
The beautiful sights through our eye anew."

"But the widespread Watchtower Zines
Always pronounce that the biological Designs

Were created by Me instead of by 'chance'!

"Just look at these eyeballs—take a glance—
And the optic system hanging behind them!
How could that come about by 'chance', these gems?"

"You, like your followers, may listen,
But You do not hear, writing with untruth's pen.
IDers deceive by this wrong approach,
Whether they mean to or not; I give reproach.

"'Chance' is not the opposite of Nature's design;
Evolution of the Species through the graduality
Of Natural Selection is the path to complexity;
Your ploy falls as flat as an imaginary line.

"A flatworm has but an optical system's spark
That can only sense but light and dark;
Thus it sees no image, not even a part;

"Whereas Nautilus has a 'pinhole camera' eye
About as good as half a human eye
That sees but very blurry shapes;
Thus these are examples of intermediate stages.

"'Rome' can not be built in a day by 'chance';
'Chance' is not a likely designer at all!

"Really now, could a 747 ever be
Assembled by a hurricane blowing free
Through Boeing's warehouse of all the parts?
Now is this the sum of Your conversational art?"

"No, Austin—it's quite unlikely—'tis just to confuse,
And that's why we always so misleadingly use
The 747 argument as the contrast to ID...

"So then, Austie, 'chance' and Intelligent Design
Are not the two candidate solutions we'll find
To the riddle posed by the improbable?
It's not like a jackpot or nothing at all?"

"'God', Your ID ideas persist, as repetition,
But again, 'chance', for one, is not a solution
To the highly improbable situated Nature,
And no sane anti-creationist, for sure,

Ever said that it was; your tale is impure.

"Intelligent Design, is neither a solution—
Because it raises a much bigger question
Than it solves, as You will soon see, in a lesson."

"Well, I'll be darned," replied the Designer.
"Natural selection is a good answer;

"It is a very long and summative process,
One which breaks up the problem's mess
Of improbability into smaller pieces, less,
Each of which is only slightly improbable,

"But not prohibitively so, thus it's reasonable,
As the product of all the little steps of which
Would be far beyond the reach of chance—it's rich!

"The creationists have been looking askance,
Seeing only the end product, perchance,
Thinking of it as a single event of chance,
Never even understanding
The great power of accumulation.

"Such they didn't know much else—their fall,
Not having any other natural ideas at all,
So they outright claimed that ID did it, as the Tree
That can magically grow the All, namely Me."

"So 'God' You have now seen the light
Of the accumulative power's might;
This is the elegance of Evolution's 'sight'."

"Yes but what is to become of Me, the Person,
For I only 'exist' through their speculation.

"In fact, the improbability of Me is so High,
And so much more so from where I lie so 'sure',
Compared to that of 'simple' Nature,
That My own origin..."

"...Is a near-infinitely Larger dilemma, Mate,
For the creationists—the problem they love to hate;
That being that You, therefore, can only be explained
By another, Higher Intelligent Designer claimed!

"Far from terminating the endless regress,
They've aggravated it with a vengeance
That is way beyond repair or redress—
As beyond could ever be yonder of! Out west!"

With that, the poor Guy faded toward oblivion,
Which remarkably was the very location
I was visiting, but hence he soon reappeared,
Although in another guise, but quite well attired.

[God created Adam, then Eve, of Adam's rib,
Both fully formed, imbued with God's knowledge
And memories of times that never were,
Such as childhood.]

[They believed a shifty talking snake,
Ate the verboten fruit,
And were cast out, to fend for themselves,
God being quite surprised at their sin...]

The God of Irreducible Complexity

*"Hello, Austino; it's time for more perplexity,
For I am now the God of Irreducible Complexity."*

"That you are, being the unmade All,
And so it shall become your downfall."

"Eh? I'm never to be at all?"

"Your believers have given You some fine new clothes:
But Intelligent Design is falsely based, God knows,
On Irreducible Complexity—
So I still recognize You as the God of ID."

"That I am is what I really am now."

"Well, Darwin said long ago that his theory
Would break down if Irreducible Complexity
Were shown to be true, and yet
No proposal has ever stood up to the analysis."

*"Still, here I am, Mr. A, alive merely by possibility,
Myself indeed quite complex, even irreducibly,*

"For "I am the be all and end all—the Prime Maker,
And so I keep tabs on every form and splinter
Of the Universe, planning its every constituent
That I designed. So then, simple I am NOT.

"Yes, man, I am an extremely complicated System,
Yet I have no parts, for then My parts that stemmed
Would be even more fundamental than Me!"

"Yes, 'God', if You existed you would surely be
Very very very complex, irreducibly so..."

"...So..."

"...So, by the Creationist Theory, such as it must be,
You cannot be explained except by a larger ID."

"I'm falling..."

"...Into the hole that they dug for you."

The God of the Gaps

Yet another Theity appeared, out of the mist.

"I am the God of the Gaps, of all those missed.
I Myself personally fill in all the gaps withstanding,
In the present-day knowledge of non understanding,

"Albeit a very large and unwarranted assumption,
But I surely do fill them all in—via the fiat lent
To Me by the creationist's fine endorsement."

"These gaps shrink as science advances anew."

"And so there is less and less for Me to do."

"What worries me is not so much that You
May be eventually laid off, having nothing to do,
But that those of Religion think it is a virtue
To be satisfied with not understanding a quandary;
Enigmas drive scientists on—they exult in mystery."

"True, My believers exult in mystery
Remaining as mystery and so they go no further,

But it keeps Me from being history!
They worship all these evolutionary gaps as being Me."

"With no justification?"

"We have a 'get out of jail free' card—a vocation;
It's an immunity to
The rigorous proofs of science;
We just claim by the 'say so'.
All must respect that stance."

"You lead a charmed life then,
One with no faults,
But You seek ignorance
In order to claim victory by default,
As a weed thriving in the gaps
Of science's fertile fields.

"Scientists rejoice in (temporary)
Uncertain yields,
Whereas You halt all inquiry."

"I remain as a mystery."

"You're the same God
Of Intelligent Design assumed—
Now known by a much more
Desperate nom de plume."

"I repeat that I intervene
To fill the evolutionary gap.
I even alter DNA."

"We could check the evidence for that.
We researchers fill the gaps in the fossil record."

"Then there are twice as many gaps. Absurd."

"I'd laugh, but I know You're not joking."

"No joke. Try what we've been smoking.
Lack of 100% complete documentation
Of Evolution means that I aid its motion."

"'God', that is not a good default stance."

"It's an unknown happenstance."

"So do we let criminals go
Because we don't have a video
Of their every intermediate foot step
To and from the lawless event?"

*"No, of course not, but we now have great worry
About our precariously perched gappy theory.*

*"Also, you made a typo—it's a <u>God</u> default stance,
Certified by nothing more than proclamation
Of Our Bull of Decree covering all instantiation."*

"An edict, huh."

"Why not, duh."

"It was also once avowed that an Evil Spirit,
One that You Yourself allowed to exist,
Produced physical illnesses, on us weighing,
But, thank God—just an old saying—
That scientists persevered, and still do,

"Such as finding out the immune system's zoo—
Our defense against the non evil spirits
Of germs, viruses, and bacterial fits."

*"Yes, agreed; that claim was dead wrong; take pills,
But evil spirits still cause the nonphysical mental ills
That are called sins and bad thoughts,
Even crimes of wills."*

"Still trying to halt scientific inquiry,
I see, for the burning.
Mental lapsing 'sins'
Stem from upbringing, wrong learning,

"And/or low serotonin and
Such imbalances, needing cures,
Not to mention the differences in cultures,

"Such as other religions
Causing a problem of stability,
For people think this undermines
Their own belief's credibility."

"Okay, I give up for now, AustinTorn. Be.
Go on with your work, with My blessing,
To discover important truths about reality,
But some fossils are evidently missing!"

"Only a tiny fraction of corpses fossilize;
However, not even a single fossil guy
Has shown up in the wrong geological stratum;
How's that for absolutely no erratum?"

"Well... it's sad for Me, but true.
I'd still love to find wrong a few,
Like a fossil rabbit in the Precambrian.
I'd have planted one there if I existed then."

"Dream on. Lazy reasoning is all that's behind
These declarations of the irreducible complexity kind."

"Yes, but all this ignorance, for sure,
Of the possible steps of Nature
Has kept Me forever alive,
Allowing Me to ever thrive."

"And has just as soon forgotten You, in truth,
But for those sustaining your being without proof."

"Wait, what about an arch of bricks?
I'll try to use this one as a trick.
"Pull one away and the arch falls apart;
It cannot survive the subtraction of a part,
So how then was it built in the first place?
With this insight, I can win the human race."

"By scaffolding, the same as seen in Evolution."

"I was afraid that would be the solution."

With that, the holely God of the Gaps separated
And nearly evaporated
To become a discontinuity Himself,
But the creationists gave Him help
By trying to hold Him together
With their ditch efforts.

(Yes, 'gapping' still goes on, it seems.

When the argument first gathered steam,
There were but a few transitional forms known,
Although good ones, enough for the idea to own,

One being the bridge to vertebrates
And another the bridge to flying creatures.
But there are many more now, a wide range,
So then it is the data that has changed.

These 'gap' arguments were already down
To the faint hope that scientists, as clowns,
Wouldn't find any more natural explanations;
But the finds were the most inevitable situations.

Creationists yet remain at the pointward
Of not being able to 'push forward',
So all that's left to is push backward,

Albeit at the firmly established fact words
Of evolution. Even the Pope concedes this
But tries to salvage the faith and solve,
By saying that the mind was not at all involved.)

"In the darkness I alit from the Wiz,
And tried to make sense of this world of His.
Now I've found the answer to life's dark quiz:
One must live this life by what light there is."

The Deity

Another God appeared, a mere Deity,
Meaning no intervention, so He's not a Theity,
And thusly said, *"Forget the Theity solution.
I am the Smart God who seeded Evolution.*

*"It was I that set the whole universal notion
And all of life's evolution into motion;
That was My elegant and foreseeing way
Of creating the kind of life that would stay."*

"I thought You were all powerful;
Why not just make 20-40 million species,
All fully formed, as immutable as Thee,
Along with their usable natural habitats,

"For this is how most Gods would do it.
What energy loss could that be to You?
Your infinity could all this in an instant do."

"I'm not so Great, plus, since Evolution is too stable
For some creationists to scoff at, as a fable,
They have assigned the job to Me, the Creator,
As all of Nature's natural Instigator,

"Because they must take retreat from the first ID God
Who zooms souls into humans at birth—it's so odd.
So, now I am not a Theity any more of proof,
And thus I must ever remain aloof.

"Of course, now I have very little to do,
And so I am not much needed, true,
For I can't even muddle with their lives;
They are all stuck now with their wives.

"I might really just as well retire,
For I am superfluous and tired."

"Well, You're still kind of close to our Universe,
Not completely outside it, the place the worst,
As I suppose your successor will have to be placed,
Absolutely, totally invisible to the human race.

"At least You made some
Basic primordial substance,
And foresaw the billion years
Of combinatorial chance,

"Predicting every turn,
Or at least knowing that something neat
Might probably come out of it,
Which was still quite a feat."

"Thank you, but it was nothing."

"On the contrary—I say verily—
You're <u>the</u> Super Scientist,
An Engineer Par Excellence—
The Ultimate Inventor of All Time—
Much better than than the old God of ID."

"Yes, I am a Scientist, making all that's real—

I Had to be, but it was really no big deal."

"You're too modest."

"It was just some little quarks,
And some electrons that I sparked,
And some forces that arose,
As reality was composed."

"But look what became of its simplicity—
Through its stages, to astounding complexity,
Over billions of years of circumstances;
We've traced the composites to simple substances."

"Well, um, it did really take that long for My intention,
By some coincidence, the same as that for evolution;

"However, I guess I'm just as surprised as you, frown,
That when some examine substance and get down
To these simple subatomic levels of unadorned things,

"That they then take a giant leap back, of all things,
To the composite complexity of Me, the Ultimate."

"Isn't complexity a much higher product
Of combination upon combination,
And thus not lower than simplicity itself?"

"Yes, it would seem so; that's a near empty shelf."

"Then I suppose You're some Great Alien Scientist, odd,
Highly evolved from somewhere, but not really God."

"True, and you, Austin, as a scientist,
Should seek what underlies the all,
Not some Great Complexity who oversees it,
For that's for what the theory calls."

"Wise thoughts."

"The best that can't be bought."

"Well, whatever on the alien thing of it,
But the creationists are not keen on scientists,
For scientists regard the honest seeking after truth

As a supreme virtue beyond all reproof.

"If they ever found out..."

"Yikes, they know not what they have made Me.
As a Scientist Myself, I truly value honesty
And skepticism over the dishonestly faked beliefs,
Those that only seem to bring Rolaid's relief."

"The Founding Fathers of America liked You,
Although some of them, as Thomas Jefferson too
Were outright non theists, many seeing You as a Deity
Who just started things up, never interfering with reality."

"Funny how President Bush's America sings,
Straying so oppositely from its humble beginnings."

"Not to mention that some the world's peoples, really,
Are squandering their precious time
Worshiping a Theity, and sacrificing to Him,
Begging, fighting, and dying for Him,

"Even threatening the world with its destruction."

"What a waste."

"Are you real?"

"No, I am but a figment of imagination, see,
But some really do like harmless old Me."

"So what's really fundamental?"
"The real fundamentals, just below
What you now call 'fundamentality',
Have always existed—the quantum reality."

"There's perhaps no time of 'forever'
At that level for Your 'always' ever."

"True, they just are, and had to be—the possible,
For a state of absolute nothing is indeed impossible."

The God of the Agnostics

I came next upon a God sitting on a high fence,

And waved to Him, saying
"Come down and talk the whence."

"I can't; I am stuck here, but Salutations to you.
I am the God of Agnosticism, one neither false nor true.
None of the agnostics know if I exist or not,
So here I must stay put a lot,

"Along with the Tooth Fairy,
Santa Claus, and the Easter Bunny,
Just in case we all might exist or not,
As a quadzillion-to-one shot."

"Why can't agnostics make up their minds?"

"My followers cannot even make or see
Probability judgments about the question of Me.
This is the limitation of agnosticism,

"Perhaps the error of no consideration
Of the likelihood of that for which evidence seeable
Is not even the least bit conceivable."

"It is a fallacy; what I call the poverty of agnosticism,
Because though being agnostic is reasonable criticism
For some things, such as whether life exists elsewhere,
It is not appropriate for those things undoable,

"For which the idea of evidence is not even applicable;
However, actually, we <u>can</u> actually still talk
About the probability of the event,
While even going for a walk.

"The true fallacy, however, is that the existence ever,
And the nonexistence of You never,
Are not even on an even footing to begin with.
The two are not at all equiprobable cases.

"The burden of proof lies with the believers,
For anything that we can conceive of
Can be claimed to exist, as that we love,
Such as ghosts, spirits, and such forth.

"Are we then to straddle a fence that has no worth?
And, never seen. So, then, at the end of the day,

"Probability creeps into the beliefs of the agnostic way,
For in practice they end up in the lurch,
Not going 'half the time' to Church,
But mostly deciding not to go at all."

"Yes, they still decide that which is 'undecidable',
For the fence is very uncomfortable
And so then the superposition

"Decoheres into the inclination
Of non belief—until, right here,
The Extraordinary's evidence appears."

He came down off the fence,
For he couldn't exist and not exist at the same time.
I continued on through the undulating hills.

(We can refer to the fence sitters as non theists
In order to get away from labels like 'agnostic'
Which might imply that the probability of thinking
God or not is on some kind of equal footing;

Plus that the fence sitters don't really stay
On the uncomfortable fence but usually...

Go one way or the other way
In life's practice of the everyday,
Although some might go to church
On alternating Sundays.

In between, perhaps they go
On wild picnics with their sweetie
And drink wine and do all that 'bad' stuff,
That we can't say here, while waiting for some
Extraordinary evidence to appear.

I will soon have a talk with
Old Jehovah Yahweh's Thee.
He's not so terrible as many
Have made Him up to be,
But then again He's not
So great either—He's quite off,
Just another poor middle manager
Caught up in the layoffs.

I already spoke to the Deity

The God who doesn't ever interfere
In the running of the universe.

The Pope doesn't know it here,
But a Deity is what he's
Leaning toward when he says then
That evolution is acceptable now
For Catholics to believe in (no mind).

The Deity Guy was
Actually kind of a great scientist.
And I already met with
The Creationist's ID God,
Who while still a Designer
Is, well, not so cool at all, either,

For He gets back to what
The Fundamentalists believe,
And neither, they would say,
Did evolution happen,
Or if it did ever function,
God constantly stepped in
To rectify its direction.

I haven't really begun
To scratch the surface of all the Gods,
Though, for so many lie now beneath the sod.

I'm only interested in
The person-type Gods of monotheism,
And I'm hardly even getting
Through those variant theisms
That fight amongst themselves
Over Jesus' divinity, or if there is a Hell,

Or a Devil and some Angels about thee,
And over so many more
And other major differences, totally.

Then there are the multiple Gods,
Now up in the millions,
And the many Gods-who-are-not-persons,
Plus the TAO, the Consciousness,
And some way-out Ones.

There are also hundreds
Of long gone, 'sure thing' Gods,
Which I needn't get into,
Except to wonder, and say:
Is that how the future will
Look at our Gods of today?

I can also skip the many
Weird offshoots that persist,
Like those saying that
The self is not allowed to exist,
Even calling it 'ego' to make
It seem so much worse;
I don't have time for these
And other cult-level verse.)

The God of the Old Testament

Of all my rotten luck,
The God of the Old Testament
Appeared and proclaimed,
"I am Yahweh, never absent,
For those schooled from infancy
In My strange ways
Have become desensitized
To My horrific side,

"And so they continue to
Keep Me very much alive,
Through their thoughts;
So, fire away at Me;
I no longer bite that hard, you see."

"You're too easy of a target to attack for free—
So it would be rather unfair of me."

"True, and I won't deny it—
It's all there in the Testament.
I was the most unpleasant character
That anyone ever made up in literary fiction.

"I was revealed to be jealous and proud of it,
Petty, unjust, controlling, vindictive,
An ethic cleanser, genocidal, infanticidal,

"Filicidal, pestilential, megalomaniacal,
Homophobic, misogynistic, sadomasochistic,
And much more, and a Bully—who gave it
Free will only if it matched My own Will."

"Peace be with you.
How about the New Testament
To replace and hide Your scent,
As many religions have already
Done through Jesus sent?"

"Yes, that Testament is quite opposite in tone,
But I am still the Father of Jesus sown,
So the problem of Me can never really go away.
I am what I was, still here unto the present day."

"Well, so long. You're the worst role model yet
That human mammals have ever dreamed up.
Who would imitate, emulate,
Or follow You as a 'leader'?"

"Well, My followers are those numerous slaves
Who excuse my mysterious [insane] ways,
Along with my exclusive desert tribe."

"Well, You're the Boss, and, anyway,
Who ever said that a God
Had to be perfect and good?"

"Everyone that I told—
And those who thought I should."

"Oh well, never mind; whatever pleases.
So, um, Joseph was not
The biological father of Jesus?"

"No, I was."

"So Jesus really did descend from David?"

"That was on his mother's side."

"Well, my ancestors descended from the trees.
Hey, why don't Catholics get the 72 virgins
That Islam gives for martyrdom for their sins?"

"I told each religious faith a different story."

"You also gave a bible half-different
To the Mormon founder,
Joseph Smith, finely engraved
On golden plates he discovered?"

"Sure. I thought at the time 'why not'."

"You had Islam add different things
To their Koran as well?"

"Yes of the many more ways to avoid Hell."

"And You told only the Catholics
That there were umpteen levels of angels
And that bread was your body
And that wine was your blood?"

*"Yep, I told just them and a few other selves,
But they made up the Saints themselves."*

"And You presented differing visions
To the Lutherans,
The Episcopals, and the Jewish,
And to many other also-rans?"

*"Pretty much,
Except that a King of England
Founded the Episcopals—
The Anglicans, of course,
Since his own religion
Wouldn't give him a divorce."*

"And you killed everyone but Noah
And his family in the Great Flood, wet,
Even young children and their pets?"

*"Sure, again, why not? Life is cheap.
However, My creation of the rainbow
Says that I'll never be so cruel again.
What can I say—I goofed. My sin."*

"But You are infallible, and even omniscient
And so You know all of the future meant.
You even broke your own commandments!"

"My omnipotence of changing my mind
Got in the way."

"But your omniscience knew you would...
One day."

"Yeah, I know—it's a paradox; oh the strife.
And I can still technically end all life,
By means other than a flood."

"You burned people in Hell, not saved,
When they didn't follow
The unfree will that you gave?"

"Yes, because I was not a loving God."

"Well, God, who made You?"

"No problem—either I was Eternal or I made Myself"

"This is remarkably the same, but for Thee,
As the Universal ingredients would be."

"Then who would need me—wait,
I don't want the answer told."

"Is the Earth only about 4000 years old?"

"Of course not but I may have let that slip to some,
To tease their intelligence apart from being dumb."

"Do you mind-read
The thoughts of every human,
Using all of your acumen,
And write the earthly script for each event,
Being so omnipresent?"

"I tried that at first, but it didn't work for Me
To put my finger on every atom that be,
To micromanage its doings for all of thee."

"That's called 'God's Will',
By some, even now.
What went wrong?
Was it the where and how?"

*"It disrupted the atoms' normal
And natural movements."*

"And that's what caused the storms unfocused,
The lightning bolts and the plagues of locusts?"

"Yes, so I stopped making such a mess of things."

"So the prayers of six million Jews pleaded
In the holocaust went all unheeded?"

*"Yes, plus I have better things to do, in time,
My sooth,
Than look after some old experiment of Mine
From my misspent youth."*

"Did you really make Adam and Eve
And all of Earth and Nature, as we believe?"

*"Yes, I made Nature,
Including the humans, in My image."*

"It shows in their rage."

"Thank you."

"God, it's ID deja-vu all over again—
I really have to move on."

*"No, wait. I like your questions.
I'm mellower now, this being My new direction.
Not as many strictly admit to Me anymore."*

"How come so many of the gospels were omitted
From the New Catholic Testament,
Like those of Thomas, Peter, Nicodemus,
Philip, Bartholomew, and more,

"As well as whole books kept from us,
Although You told some other religions to keep them,
Such as the Book of Revelations?"

*"Those gospels were embarrassing and wild;
They told about My Son doing magic tricks
And practical jokes on people when He was a child."*

"Oh, we never heard much about his youth.
And didn't You send the Mormons proof
That Jesus spent an early era
In what was to become America?"
"Probably."

"What about the trillions of galaxies in the sky?"

"They're just for show and scenery on high."
"Where's all your rantings and ravings
That I've heard about?"

*"I now take Prozac for
My mood swings and bouts."*

"You don't really exist, do You, as mental,
For how could You have an emotional system—
As composite—and still be absolute and fundamental?"

*"No, I don't exist,
For how could I since I am so horrible?
Human mammals made all of Me up
As a very bad example,*

*"As it turned out, from their many fears
In the childhood of their species' years.
Unfortunately, it caught on to their children's ears."*

"So, yet You still subsist
In this indefinite locus of wishes?"

*"Yes, sort of.
I am sustained here since many children
Have learned to obey and listen
To what is-was told to them,*

*"For this obeying was an
Evolutionarily useful thing,
As many of their obediences
Resulted from warnings of things*

*"That were truly dangerous,
And so the children grew up
To indoctrinate their own children
In all the 'knowledge'."*

"We'll have to offer more reason
To those so indoctrinated.
Now farewell to You, the impersonated."

*"See you. Pay no attention to Me as certain,
But to all those blinded by the curtain."*

He soon dozed off into never land.

The Gods Meet One Another

I next encountered all the individual Ones,
The specialized Gods of all the Religions.

They didn't get along at all, not even for an instant,
For all they had in common was their intolerance
Of the others' greatly erroneous and unjustifiable beliefs
That clashed with their own, for tolerance as a relief
Was truly Not an attitude the jealous Gods endorsed.

The followers of each God thought that their own
Irrational embrace of myth trumped the others' known,
And so this led to many of the religious groans.

I watched the Gods battling for a while, steadfast,
In the present, as well as in the distant past,
Their followers' beliefs scripting the actions,
Conflicts leading to dying for untestable propositions
About where everyone came from and was going to:

Metaphysical Martyrdoms
Conflicted with the Divine Book of Revelations.

Deuteronomy 13:7-11
Stoned those disbelieving in Yahweh,
Killing them, while the Koran eliminated some infidels.

India and Pakistan, different countries domiciled,
Because the beliefs of Islam could not be reconciled
With those of Hinduism, were poised at the brink
Of nuclear war merely because they disagreed, rife,
Over some supernatural 'facts' concerning the afterlife.

Karmas ran over Dogmas.

Musharraf suspended Pakistan's constitution,
To stamp out the growing Islamic militant coalition.

Palestine's Jews and Muslims scuffled on;
Balkan Orthodox Serbians dueled
With the Catholic Croatians,
As well as with the Bosnian
Albanian Muslims;

Northern Ireland Protestants
Warred with Catholics;

Sudan Muslims discorded
With the Christians;
Sri Lankas's Sinhalese Buddhists
Went against the Tamil Hindus;

Caucasus Orthodox Russians
And Chechen Muslims
Exterminated each other and their kin;

Iraq's Sunnis and Shites massacred each other
For some very slight dogmatic differences.

I interrupted their skirmishing and said in haste,
"What about tolerance and respect for other faiths?"

They all answered at once and said, in unison's beef,

"That's just political talk. If we tolerated other beliefs,
That would be akin to recognizing them readily
As having some credibility, which they certainly do not.
We are saved and they are all doomed, in peril;
We can't have them exerting influence in the world."

"So," I said, trying to make some small talk,
"I've heard that You've each written a book
That makes an exclusive claim as to its infallibility.

"Congratulations to each of You on being published.
All have made the bestseller list;
However, I have respectfully shelved all of them
Next to the *Egyptian Book of the Dead*
And Ovid's *Metamorphoses,*
In the contradictory book and bible section.

"Hey, how about getting modern and making a film?
I know that a book was a great thing way back,
But a moving picture is worth 10,000 still pictures
Which are in turn each worth a thousand words."

*"Indeed, we will each be divinely inspiring a movie
That will soon be playing in a theater near you."*

"Wait, Guys, I take it back," I said with alarm,
"Are not all your children doing enough harm
By fighting over your books and morality plays?

"Will people now die for another media—the movies?"
They ignored me and fought on, with their kind,
Unable to see but through their own 'right' minds,
Doing the opposite of their teachings of love,
Which they were especially and paradoxically out of."

Unfortunately, they now represented the largest threat
That human kind has ever imposed against itself—

All due to differences regarding some very improbable
And differing notions about the nature of the universe.

I noted the Land of Evil Demons,
Although sometimes it was hard to tell which
Was which or not witch.

I also bypassed the numerous Gods
Of the instant Cults
That had always gained so many followers,
And bad results.

God of the Religious Moderates

I next encountered the
God of the Religious Moderates,
Whose numbers had been
swelling lately, at any rate,
But they had seemed to
get stuck in that middle state.

The God of Moderates said to me, in soft oration,

"Greetings. All things in moderation."

"I bet that You derive from secular knowledge
Combined with religious ignorance."

*"Well yes, modernity has allowed some dust to settle
On the very old unchangeables that do nettle,
And so now people pick and choose,
Invent, or ignore the Dogma's ruse."*
"Dogma is indeed
An unchangeable definition—
It does not admit of progress,
By its very definition."

*"True, but I am still their God, of course,
As they have abandoned the wingèd horse,
Virgin births, sexual prohibitions, the value of life—
And they even have some doubts about the afterlife."*

"They betray both faith and reason."

"That they do in this new season."

The God of Nature

Lastly I met the God of Einstein—Spinoza's God.

*"I am the so-called God of Nature,
Being as one and the same with it—no different;
Although that which has no difference
Is really not any different.*

*"Anyway, at least this is how the people awed
By Nature's intricacy and beauty refer to Me.*

*"I am only here in this nebulous vicinity
Because I don't actually exist with certainty,
But seem to some to be tautological with Nature,
Always existent and beautific."*

"It's OK, don't worry about it."

"Thank you, and welcome to reality."

"You mean I'm back?"

"Well at least you have one foot in it through,
Just as I seem to do."

"I'm going, but why did humans
Invent the theistic and deistic Gods?"

"Man created them in
His image's inward glance,
Because he was and
Is terrified of his insignificance,
As well as from a fear
Of losing the beauty
Of his life's instance."

"So man just proudly declared
That he was of Special Creation."

"Yes."

"Farewell and thank You for Your insight."

He called after me.

"Enjoy reality—it's really a place that's better.
There's nothing more beyond it.
All comes from matter;

"You're electrochemical creatures—
As organic and natural
As anything else in Nature.

"Consider this knowledge
As the ultimate humility, if you will.
Live life, love it—while you can,
During your lucky incarnation
From the evolving composites
Of the last 13.75 billion years.
You are here. You have arrived."

"Panthea, the greatest God there never was...
How to explain? She does what nature does.
As a rose is still a rose by any other name,
Then so is a universe a cosmos the same."

Down to Earth

As I rejoined Actuality, I felt its waves and seas
Of brightness and color joyfully washing over me.

Getting back to my existence and its stresses,
I ignored some knocking Jehovah Witnesses
Then made nine golden tablets,

And reported my findings on ToeQuest,
Then went breathing, seeing, and hearing,
And otherwise sensing
All that was knowable as reality.

THE PROBLEMS OF TRADITIONAL RELIGION

The Christian concept of reward and punishment
Handed out by an omnipotent, omniscient God,
Is derivative of the family experience—
The child and parent—a conception of our world.

God in the News

I picked up some newspapers and magazines:

A suicide bomber blew up a bus and himself as well,
Sending many of the unbelievers straight to Hell,

While assuring himself and 72 friends a place
In Heaven, a double blessing from his Faith.

His family, relatives, and friends gathered soon
To celebrate their wonderful good fortune.

The bomber's death was especially lauded as wise,
Because he had proceeded directly to Paradise,

Bypassing the possibly troublesome way
Of the litigation of Judgment Day.

Fighting continued in Kashmir
Due to some perceived insults to Muhammad.

A man was released in Northern Ireland
After claiming to be a Protestant atheist.

A child of Christian Scientists died
Due to the religious refusal of antibiotics.

Extremists sought nuclear formulas and parts to reduce
The peril of the unbelievers in the world,
Those whose ways are not sanctioned by Allah.

Pope authorizes millions to reach
Children sexually abused by priests.

The recently discovered Gospel of Judas
Suggests he wasn't really such a bad-ass.

Some nuclear facilities no longer exist in Syria,
About whose disappearance both Syria and Israel
Seem to know nothing.

Battles rage on over differences in some holy books.

Iran promises to destroy Israel.

President Bush led off his latest speech with
'In God we trust."

And in a more than 2000 year-old newspaper:

The Emperor led off his latest speech with
'In Zeus we trust'.

And finally, in a future newspaper:

Religious extremists detonate atomic bomb
In Washington, DC;
Nuclear retaliation destroys
Twelve highly populated middle-eastern cities.

World greatly stunned, begins to widely read
The End of Faith', 'The God Delusion',
And 'god is Not Great.

THE ON-LINERS AND THEIR NATIVE LANDS,
Especially Australia

Melanie and Jacy live in a U.S. satellite,
A land of mist and not driving right,
Called England, Mel having been born there
As the Loving Goddess somewhere
On the sacred Isle of Women—
Then taking off with Antonia swimmin',
And Jacy, a woodcutter's fine daughter,
Guards the border with her wonderful laughter,
Her thoughts on the loose, gone through the rafters.

In that country, 'left' is whatever is right,
Leaving 'right' to be whatever is left. Right?
When I went over there to see her,
I always had to use a mirror.

Melanie uses both brain hemispheres,
Mostly in the pursuit of achieving there
The brainless bliss of nothingness,
As of the Bra-Man's dreaminess.

England's not as bad as Rav's, Jamtimes's, Leo's,
Lomax's, Pytor's, Graybeard's, and Tina's
Upside-down provincial dominion extrema
Of the Mars-like planet of Australia,
Which has a bunch of dry dusty towns
Where they say "What's going down?" each day,
Instead of "What's up?", but, either way,
The answer is always an elevator or a lift.

A foreign lady at a hotel once asked me for a lift,
So I picked her up and carried her—quite miffed,
All the way to her room, as a welcoming gift.

I flew down under to Australia only once,
On assignment with Bill Bryson's penal sentence.

Australia is a mostly an empty thralldom,
Much more vacant than even an atom,
And is extremely far away from anywhere else.

It has less people than Tokyo,
But is a zillion times more extremo.
The constellations are the inverse

And the seasons run in reverse.

It has nothing of any interest
And the climate is the cruelest.

It is the only country that is also a continent
That is also the world's largest island extent
And the only one begun as a prison meant;
Graybeard must have a lot of original sin
Because all of his ancestors were his criminal kin.

The cities are all on the coast since the interior place
Is an endless desert about half the size of outer space.

Somebody once set off an atomic bomb
In the Great Victoria Desert's silent calm,
In Western Australia, a land much embalmed,
But no one noticed it for years—that maelstrom,
But for creatures jumping right out of the genome.

Of the world's ten most poisonous snakes,
Most are Australian, for Christ's and God sakes!

Even a fluffy caterpillar can kill you;
Seashells are venomous too;
Adieu and skidoo to you!

Every ocean current carries you far out to sea.
They even lost a prime minister, Harold Holt, see,
Who was merely strolling along the beach.
He stepped into the surf, going swiftly out of reach,
And was never seen or heard from again, 'imbeached'.

Australia is very old and nothing has changed
There for 60 million years, nor anything rearranged;
Thus one may find there the oldest fossils on Earth.
Even the first faint signs of life can be seen: its birth,
And the earliest animal tracks ever made; no dearth.

It seems, though, that its creatures evolved
Outside of Darwin's book, being quite convolved;
They don't run at all, but just bounce
Across the landscape, like a ball, or they jounce.

Australia is the driest, hottest,
Most useless, infertile, flattest,

And climatically unbalanced
Of all the continents instanced.

It is so hot there that recently, amen,
The air caught on fire once again.

The place is so inert that even the soil is a fossil;
Even the worms and bacteria are quite docile.

On the up side, they have 20%
Of the world's slot machines present
To serve less than 1%
Of the world's population extent.

As about equal to finding a live T-Rex,
Proto ants were found alive there, having sex,
Although nothing like them had existed on Earth
For over a hundred million year's worth.

So does Graybeard know his evolutionary stuff
Or what? Yes, indeed, plus all of those ants so tough
Were found on his back porch, hellbent with intent.
They've now become extinct due to his experiments.

I flew to Los Angeles from Australia, getting there,
In time and date, even before I left the shore,
Which was hardly soon enough for me;
Let us all say a prayer for poor Graybeard to be.

The Science News is that NASA is feverishly pursuing
The manned Mars Landing plan, fliers ever wooing,
But still needs some suitable astronauts,
So it's off to White Cliffs Australia for some kumquats.

Here they found a population of about 80,
In a wilted world of heat, rocks, and dust, matey.
Due to the horrible heat, their cave-houses
Are burrowed into the hills—for souses and mouses.

When a vehicle goes by, it raises a big cloud
Of red dust that eventually settles, enshrouds,
And covers everything in sight, leaving nothing to see.

They've had electricity only since 1993,
TV since 1998, but no channels yet to espy.
Taipan snakes slither by, on the sly,

Their venom 50 times more poisonous
Than a cobra's—to leave you breathless.

Australia began as a nation that was thrust
When convicts actually began wanting so much
To go there for the crazy gold rush.

So anyway, NASA enlisted them all
In the Mars Landing Program's shortfall,
Figuring that they wouldn't really know the diff
Between the planet Mars and White Cliffs,
Or that if they ever did they'd be so spaced—
Happy to reside in a more hospitable place.

Science Saves Us From the Warm Future:

The White Cliffs Underground Motel moochers
And those of the various home residences picked
Have the right idea to avoid the warming conflict:
Free cooling to 67 degrees F: perfect!

This may be a good plan for all of the future deal
If global warming really happens to happen for real.

I can imagine a stay in the Dug-Out Motel, unraveling,
It being quite a heavenly destination after traveling
Forever, going over bumpy roads and then getting
Out of the 'blender' mode and into the pool, wetting,
As all this traveling would grant more appreciation
Of the three star AAA motel's accommodations.

Coming, sweetie?
(Just you and me.)

Our room would have natural light from a shaft,
Which saves on electricity, oil, and gas.
There are no windows, but that only saves us
From having to view the non-scenery—a plus.
Cell phones wouldn't work, but hey, happy endings
Then for those blendings never tending;
There would be no interruptions
Of any pending eruptions.

Hey, how come people in soap operas
Always answer their darn phone; sagas?
It always ruins the moments erotica.

Plus we could always dine in the restaurant,
Since the nearest supermarket is a scant
And rough six-mile drive away, askant.

And those dust-assisted sunsets
Are of truly unbeatable descents,
They having ten times as more
Colors than the rainbow: fourscore.

Plus with White Cliffs having electricity now,
The beer tastes no longer like a steamy hot cow,
At 110 degrees F, but ice-cold, for highbrows.

There was a bad drought in the 1890's here
And the land has not recovered, oh dear,
But who needs that when one has love and cold beer.

And now of others whom I met online,
Gently roasted here with rhythm and rhyme.

In Fred's pyramidal world, it is that the origin
Of the belly button of the universe was an 'inner' begin
That reached the limit of being really small,
And so it popped back out to become the 'outer' all,
As in the outer space of the entire universe around
That formed from the unlimited merry-go-round.

Mohan wrote many multi-verses of poems
On the steamy planet of India, per diem,
Where they have three million Gods become.
They will eventually be getting more,
So that each person can have one to adore.

One day the temperature there went down
To 75 degrees F, which would be a perfect markdown
Anywhere else, but here they all tried
To look for sweaters to put on outside,
But they didn't own any,
At least not very many.

In the summer, on and in the icy planetoid
Called the Yukon, a frozen place that's best to avoid,
LabelWench worked only at night,
But like the day it was just as bright.

The sun never rose, staying up all the more
For it had never set during the day before.

When a cloud came by in its starkness
They called it night or darkness
And had to use flashlights until it passed
So that they could all find their wine glasses.

They have only two days a year there,
Each six months long and longer,
Called white and black
Or bright and pitch black.

MJA lives in a land where there was no
Difference in anything, for all was equal.
Everyone was a clone, wearing the same clothes.
The sports results were always ties—so close.

Everyone got an 'A' in school,
For they couldn't measure the rule;
But they'd all progressed beyond this equality,
Thanks to Bottomlander, who lives in a valley.

Antonio lives in the celestial body of Mexico,
But for some reason they call it Texas, since long ago,
The U.S. stole it away. We would give it all back
But for the fact that they have already taken it, Jack.

I always try to say 'Remember the...', the start,
But I usually can't recall the 'Alamo' part.

One time I got a letter, for my laxes,
For Austin, on taxes
From Austin, Texas
And I didn't know what to do but axe it.

In Greenbug's asteroid of Greenland,
Every single thing was green, and,
So after a while this gave him the blues,
After which valley of depression's dews
He then felt very much in the pink,
And much on the uprise until all was a rosy think.

Then he discovered that he had been given
Green contact lenses at birth, these being riven;
His land was really all white,

The 'green' in the land name's write
Being only part of an advertising plan
To get people to settle the land
There, after not so many came to Iceland.

Bogie resides in the sunken land of Florida
Where the year-round heat is all too horriba,
Where old people walk really slow in front of you
Towards God's waiting room, to the very last pew.

Bogie cools his thoughts in the arena of Tampa Bay,
Pondering every idea that comes his way.

RascalPuff lives in Niihau, Hawaii,
A secret place; so that's all can say I.

Graham lives in the Canadian clouds
Where all is allowed,
In a levitated home,
Smoking pot homegrown.

Felix lives in Schrodinger's cat-house shed,
But only half the time, when he's not dead.

Lloyd lives in the real house of science,
So please let all posts there be in compliance.

Leskey's leaving the new land of Zeal,
Ever becoming more and more real.

Max lives in the U.S. in the state of Deep Thought,
With all his relatives, many of whom were fought—
Cousins twice removed, but they kept on coming back.

Melanie says that "Nothing is Real"
And this reminds me of guy whose spiel
Related that "Nothing is true". For real!

Everyone believed him for 20 million years
But then they found out he was lying—oh, tears,
And so it was then so sure that they truly knew
That the case was really that nothing was true.

And so for 15 million more years of bluffing
They believed not anything and nothing.

So what was this guy's name?
Well it was the man with no name,
Which was Nobody Nowhere
Who was now here but no where.
He lived in the Noplace
Of virtual space.

Then a large but tiny problem
Was found with nothing, ahem,
That there was a slight ado about it, surely,
This being the quantum uncertainty,
Not a very big deal, really,
Being the smallest thing of reality,
But enough to raise it to be a near nothing,
Just about as close to nothing
As one could ever get, without stuffings,
And so it hardly really counted for much,
But it made for a universe in which, as a crutch,
The gravitational energy was negative,
It canceling out the positive—
All the energy of stuff,
But for the unavoidable touch
Of the quantum uncertainty
Which we can almost certainly
Avoid for all practices, purposely.

So, "Nothing is true" and "Nothing is real"
Turned out to be pretty much right, a done deal,
Except in England, where it was all that was left:
Reality bereft, a cleft from the theft that was deft.

Meanwhile, Mel shivered with the quantum jitters,
Turning it into a jazz dance of some random twitters.

Now, what about "Nothing is real",
Employing it in the sense that the real
Doesn't even exist, although it has a feel.

Well something does exist,
So we might rather say that nothing persists
And so "Everything is temporary" but being,
Since our realism came from a near 'nothing'
And to such it must return, in great arrears,
Even if that takes about $10**10**10$ years.

ProfPat is expanding into the void, accounting for

The Catholic girls' heavenly student bodies more,
And lives in the naturally divided state
Of Michigan, that unmarried state
Which is separated by a long fat lake,
A further segregation being that the upper part
Of the state associates with Canada,
While the lower part is called "Michiana".
On the other side of the lake the state
Is more or less a part of Wisconsin's fate.

Up Above:

One time I drove from Chicago's exploit
To New York by entering Canada from Detroit
Into Windsor, near the Church of ProfPat; adroit,
By then getting out of Canada's shortfalls
As soon as I could, at Viagra Falls,
A fun place for vacationing foreigners
To leave lots of money for souvenirs.

When they say that the glaciers retreated,
They only mean that they repleted
And went back up into Canada, where they sit
Atop the forgotten land, in which mitt
Few are cold because so many are frozen stiff.

Since it is all ice everyone plays hockey all day,
At least when they can blowtorch the ice away
From their igloo'd-cars, if the flames will stay.

Canada has only one super highway:
It goes east and west all day
And as close to the U.S. as it possibly can.
It doesn't even have railings, man,
For there is nothing to run into
If you go off the road by some miscue.

So anyway, Canada is only really only about
A width of ten miles that is a barely habitable shout
Just above the U.S., a suburb really, for fallout.

They have only one baseball team
That is going nowhere, it would seem,
Since they are really all hockey players extreme.

The police still ride horses there, on the loose,

But this is actually a step up from a moose.

Sears is their biggest industry, barely afloat,
Mostly selling really heavy fur coats
Made from polar bear furs that were poached.

Mikal works in men's clothes there.
She drank Canada dry in her time that was spare,
But now drinks only ice water, right from her tap,
Being sober and serious and all that yap.

All their restaurants are called Tim's Donuts.
Canada is really even smaller than it looks, a rut,
For a part of France called Quebec is in it but
Is not really with it, they all being nuts.

Every U.S. map I've ever seen ends at Canada,
Just showing it as a bit of a blank gray area,
Which probably is really the right sceneria.

When Henry Hudson discovered the frigid Hudson Bay,
His men were so mad that they put him off one day.

Mikal lives at the end of long Lake Ontario,
A prime spot since it gets the extra lake effect snow
Of two feet more than the average of five feet or so.
June is still a winter month there in that fen
But one only needs two coats to wear then.

One time it got up to 80 degrees F there
On a mid-summer's day and everyone there
Became sweating and boiling and so there
They all ran around with naked eyes all bare.

The power goes off more than it's on.
The only place worse is Antarctica wan.
Even the Yukon, which is really a secret part
Of Alaska, has better weather than Ontario's heart.

So to help Mikal, let us apply some science
In the form of some secret zapping rays of potence
From the North pole's Harp Array that will melt all the ice
And then flood Canada ten feet under; very nice.

"What's the big news from Canada, Benny?
They don't have any, Penny."

As for those in the rest of the world orientated
It is that the Easterners are dis-oriented
By the Westerners who resist any other orientation.

...

Now for some outdoor non online fun,
With no one on the computer, but out on the run:
Everyone was out having a ball lately by hitting it
With a club, bat stick, or a racquet.

Graybeard shot an eagle and a birdie
And then cooked them for a dinner fry.
SB_UK scored a wicket, whatever that means.
Arthur argued the laws out in left field
With an umpire who was always right.

LabelWench jumped her horse over a giant snowball.
Max rolled a bowling ball down his road at ten trees.
Austin avoided the [tennis] net of evil
And tried to keep within the white lines of goodness.

Mikal thought that a sand trap
Near a water hazard was a beach.

MJA's equal game ended in a scoreless tie.
Melanie scored 18 holes in one because it
Was really the Perfect Awareness that was playing.
TimeParticle hit a golf ball with a baseball bat
And created a new orbiting moonlet.

I never played polo, but I played golf,
Which I learned from playing billiards,
At least the putting part,
And so I suppose polo is really golf
Combined with riding a horse.

TimeParticle is so strong
That he hit the ball out of sight.
Melanie said that there really is no horse and ball.

Graybeard reached for a branch
Whenever the horse's ears twitched.
Graham used a levitating magnetic horse
After the bull moose threw him off.

Austin went hoarse from too many posts.
Wick played from the 4th dimension with a hypersphere.

Everyone who was dying to find out
What happens after you die
Almost died laughing and nearly found out.

ON THE ROAD OF TIME

She loves road trips. The autumn colors called,
So we were off on the ups and downs,
She with taped ankle and myself with wrist,
The warriors running away from home.

The scene was of the turning leaves falling,
Unspoken poems reciting the paths flown,
Only now the scene painted with the words,
As music played poems sung to melodies.

Country roads, quaint inns, dilapidated barns;
What's this? A dance hall lighting the dark path.
We dance the song of evening bells rung
In a twilight zone in nowhere's middle.

The music played past but not yet past,
For it was in recent memory recalled.
Newly savored sensations continued on—
Those which could be presently known.

Mind anticipated the coming tones,
The transitional 'middle' blending it
With those sounds not totally gone.

In this past-present-future resides
The delight that none could produce alone:
The smoothly rolling 'now'.

The Meanings of the Names for
The States and some Provinces

<u>Alabama</u>
"I'll a bam ya if you call me
a 'hick' even ten more times!"

<u>Alaska</u>
From "All-last-ka", since it was the last state
that we stole from the Indians.
We would have taken it sooner,
but we didn't know it was full of oil and gold.

<u>Arizona</u>
Perhaps from the O'odham
Indian word for "little spring",
since it is summer there all year long,
kind of like being in the twilight "arid zone".

<u>Arkansas</u>
"Arked we still in Kansas?"
"No, Dorothy, not anymore,
ever since the flood."

<u>California</u>
Cal-I-fornicate;
a loose state with few clothes;
looser still after the next earthquake;
best to take out no-fault insurance.

<u>Colorado</u>
From the color "ado", meaning
"much ado about ruddy or red".

<u>Connecticut</u>
A paradoxical state,
since first they "Connect",
then "I-cut."
What's it going to be, guys?

<u>Delaware</u>
Named for Sir Thomas West (huh?),
Baron De La Warr (oh).
I was not even aware of Del,
nor of the state, since it is so small.

<u>Florida</u>
"Where's the floor, Ida?"
"Um, it's under sea level now."

<u>Georgia</u>
In honor of George II of England,
for he was a cross-dresser, I suppose.

<u>Hawaii</u>
"How are ye?" (meaning 'Aloha').
Quiz: what state is an island?
Nope, wrong;
the answer is "Rhode Island".

<u>Idaho</u>
This is an embarrassing state for women:
"I da ho".

<u>Illinois</u>
"Ill-noise" or Algonquin
for the "tribe of superior men",
this being the same thing.

<u>Indiana</u>
Meaning "land of Indians",
or for a friendly squaw
endearingly known as "In-Diana".

<u>Iowa</u>
"I owe ya... something; well, forget it."
Or from the Iowa River, which was named after Iowa.

<u>Kansas</u>
From a Sioux word meaning
"people of the south wind" or "Kans-asses".

<u>Kentucky</u>
From an Iroquoian word "Ken-tah-ten",
meaning "land of tomorrow",
for it is still a very backward place.

<u>Louisiana</u>
In honor of Louis XIV of France
and his secret lover, Anna.

Maine
First used to distinguish the mainland
from the offshore islands. Not very funny.

Maryland
In honor of Henrietta Maria
(queen of Charles I of England).
No gay marriages are allowed here,
nor any practicing of marriage without a license.

Massachusetts
A sneeze in Catholic church.

Michigan
From the Indian word "Michigana",
meaning "great or large lake".
Or, after Mitch, who began the state.
It's not finished yet, either,
for it is the only state still in two pieces.

Minnesota
From a Dakota Indian word
meaning "sky-tinted water",
or a "small soda".

Mississippi
Named after Mrs. Zippi.

Missouri
Named after Miss Youri.

Montana
Named long before Hannah Montana.

Nebraska
From an Oto Indian word meaning
"flat water" or "utter wasteland".

Nevada
Spanish: "snow-capped".
Casino odds: handicapped.

New Hampshire
From the old, old, really old
English county of Hampshire. Really.

New Jersey
They always buy and wear new sports jerseys.

New Mexico
New "place of Mexitli,"
an Aztec god or leader who snuck across the border.

New York
In honor of the Duke of York's new playground.

North Carolina
In honor of Charles I of England's
wife "Carol" and his girlfriend "Lina".

North Dakota
From the Sioux tribe, meaning "Da cold, no?".

Ohio
From an Iroquoian word meaning "great river".
It is not high in the middle and round at both ends,
but is totally flat.

Oklahoma
From two Choctaw Indian words
meaning "Okla's homa tonight".

Oregon
Unknown, but all the ore is a-gone now.

Pennsylvania
"Pencil-Vania", in honor of Sylvania,
who invented the first pencil,
or, in honor of Adm. Sir William Penn,
father of William Penn,
who invented the first pen
somewhere out in "Penn's Woodland".

Penn State
Name has been changed to State Penn.

Rhode Island
From the Greek Island of Rhodes.
It is not an island, but Hawaii is.

South Carolina
In honor of Charles I of England

and the more southern parts of Carol and Lina.

South Dakota
From the Sioux tribe, meaning
"We don't get along with North Dakota"
or "so, da cold?"

Tennessee
Where one can see the best tennis players.

Texas
Stolen from Mexico,
but now it seems as they're taking it back, Tex.

Utah
From "You tall as a mountain",
or, you of certain sect
need 12 teenaged wives for certain sex.

Vermont
From the French "vert mont,"
meaning a "green mountain", of maple syrup.

Virginia
In honor of Elizabeth, "Virgin Queen" of England.
They may check you at the border
when you enter to see if you have been entered.
Then you may live happily thereafter, in Virginia,
but your wife may get jealous.

Washington
In honor of George "Washing a ton of clothes full of dirt".
He slept around a lot in any old place, it seems.

West Virginia
In honor of Elizabeth, "Virgin Queen" of England's
western, untouched regions.
Note: East Virginia blew away in a hurricane.

Wisconsin
French corruption of an Indian word whose meaning is
disputed, possibly meaning "Wish you'd come and sin".

Wyoming
From the Delaware Indian word,
meaning "mountains and valleys alternating";

the same as the Wyoming Valley in Pennsylvania.
"Why, oh Ming, did you not stay in Pennsylvania?"
"It wasn't even worth a penny."

U.S. Virgin Islands
No one lives there anymore;
all have been disqualified.

Marriage
A state of confusion.

District of Columbia (DC)
A state of confusion.

Ontario
A suburb of the U.S.

Newfoundland
Not new anymore, for the rest of Canada was much new-
er.
Named after a new found land.
Was recently blown away by hurricane Igor.

Quebec
Actually a part of France.

Alberta
Named after King Albert,
who was also a Queen.

Yukon
Can't say "Yuck on this place!"
or my friend, Scheherazade, might get mad.

THE GREENLESS WORLD

I'd come to this strange and foreign world
Over three years ago as a scout
For a phosphorus mining expedition,
And here I had remained, marooned,
Since the nearest asteroid supply bases
Had been closed for lack of
Their necessary Earth supplied material.
Well at least I had life; I'll take that anytime.

I was thankful too that my alien friend,
A native of this planet, was female,
And that we were compatible
Both genetically and physically,
Although we were probably unable
To produce offspring, at least so far.

Science long ago had indicated that the Earth
Was probably not the birthplace of mankind,
That Earth was seeded by ancestors
Who were common to all the galaxy.

My friend's name was Serena,
That being the closest English translation.

Over the years here,
I had learned her language
And she had learned mine.

We lived together 24-7,
And so I had been spared
An eternity of loneliness,
Although it had been a very close thing,
I being the only human here
And she being one of the few remaining natives
Of this large but doomed planetoid.

This planetoid had been dying since its birth,
For it had three suns, one of which was always shining,
And so it was only a matter of time, I suppose,
Before all the underground springs evaporated.

But these types of geological events
Were still measured in centuries, if not more,

And there was perhaps
No immediate danger in our lifetime,
Although life here was certainly
Becoming more difficult;
Hence the already great exodus long ago
Of those who could afford to leave.

There was never any darkness in this land
Where the sun always shone,
Not even inside the caves,
For the phosphorous in the walls and the ground
Gave off a constant luminosity.

This phosphorescent light
Had been hard to get used to at first,
Although Serena had no problem,
Having been born here;
She even had the natural ability
To sleep with her eyes open.

It was the hottest part of the week now,
The time when the two largest
Of the three suns shone at once,
There often being such an overlap
As this for days at a time,
And so we often had to retreat
To the "cool dampness" of our cave—
Our home in this primitive world.

Even when there was but one sun in the sky,
It was still quite unpleasant to be outside,
For it was always hot, and bright too,
For the suns, all of them, were large,
And one could not easily look up
Into the sky near any one of them.

The cave was lit
By the radiant glow of the walls.
No real blackness anywhere.

Our lunch was boiled brown vegetation,
The only cuisine available;
However, when one is hungry,
One is thankful for anything at all.
No gourmet food here.

Serena had never known darkness, and, indeed,
There wasn't even a word in her language
Which meant anything close to "dark", "night",
Or even "black"; however,
I'd been able to convey the concept
By using the absence of light as an analogy.
Of course she still had trouble grasping the idea
Of "that which could never be".

I suppose it was like asking someone
To visualize a color that one had never seen.
Naturally, I tried covering her eyes
To simulate darkness,
Since she couldn't close them,
But she still reported a yellowish color,
And later, upon inspecting her eyes,
I noted that they gave off a cat's eye type of glow—
Just like every darn phosphorous rock on this planet.

Even the sand shone like gleaming yellow snowflakes.
Ironically, this was what had brought me
Scouting here in the first place—
The prospect of mining that rare yellow light
That made fireflies glow
And caused those struck matches of old to light up,
For the Earth's supplies had long since run out.

This ever present light was at first
Psychologically disturbing,
But I'd learned to live with it,
First by sleeping with a band of cloth
Wrapped around my eyes, although gradually
I lost all track of time and just slept
Whenever I got tired.

I also had to be careful not
To come into any rough contact
With the phosphorous rocks in the cave walls,
Lest they should burst into flame.

Yes it was a rather precarious existence,
Though a livable one, but alas,
I could never go home again,
For the Earth had been destroyed
By a giant comet,
One of the Perseids, the shower

Whose many precursions
Had given us the wonderful
Meteor shows of that name.

I turned to Serena and spoke to her about it,
Having been unable to deal with it until now,
And also because she had only recently gained
The scientific knowledge to be able
To understand solar system concepts.

"I was one of the lucky ones, Serena,
For I was already out in space,
Had just recently launched, in fact,
When the disaster hit my home planet, Earth.

"You cannot know the shudder that went through me
When I realized that all that I had loved was gone,
That all that it was or could be, all that had formed me,
Given rise to me, was gone forever.

"My Earth was one of the most priceless work of art
That the universe had ever known.
This rock on which we now live
Is not even Earth's pale shadow—
At least we do have pale shadows on your planet,
Though they are hardly noticeable.

"At first, when I saw Earth's fireball,
I thought that I had seen a shooting star, but then,
Noting the origin and size of the spectacular explosion,
I was overcome by a horrid feeling—
One that was chill and sickly like any I'd never known—
That it was indeed that the Earth had left us.

"I could do nothing but continue on,
For the Earth had no equal in our solar system.
Oh we had long ago searched
The entire galaxy in vain for such a paradise,
But the Earth had remained unmatched."

Serena thought for awhile,
Having only recently grasped the idea
Of a universe filled with worlds,
She never even having seen stars
In this land in which night had never fallen.

But again I was fortunate that
She had an open and intellectual mind;
So during our recent studies
I had been able to take her thoughts and her mind
Across many centuries of learning and knowledge,
Sequentially educating her step by step,
Using small and primitive learning blocks
Until reaching some rather complex theories.

She was now able to understand such concepts
As solar systems, space travel, physics, biology,
And many other unseen wonders
Like oceans, rivers and lakes,
Which though quite impossible on her planetoid
Were at least conceivable to her
Since she had often seen water bubbling up
From the hot springs—which, by the way,
Were apparently the limited
And unrenewable source
Of both water and oxygen
On this near-planet.

Finally she spoke,
After allowing the cloud of sadness
To pass from my brow,
For she was emotionally very capable,
"Patrick, you lost everything that day—
You are a man without a world.
How many people died? How many survived?"

"Trillions died—that is a number you don't have here,
But take your "deca" and multiply it by itself
For "deca" number of times
And you will be close to knowing what a trillion is.
No one on the ground survived."

"A trillion is like the number of grains of sand
In the desert outside our door," she answered.

"Yes, Serena! I might of said that in the first place
But I suppose I've been too much of a scientist lately.
As for how many survived,
I'm sure that's only in the thousands—
Perhaps eventually only in the hundreds,
Since many Earth outposts were contained
Within domes on uninhabitable moons and asteroids

And were quite dependent on the Earth
In the long run for their survival."

"I understand more and more everyday,"
She answered,
For she was now quite proud
And even happy with all the ongoing revelations.

"When you first fell from the sky
I thought you a god,
But now that you explain everything
I see that it all makes sense,
And what once seemed magical and clever to me
Is now all laid bare before my eyes
As something entirely reasonable."

She spoke mostly in English now,
There not being enough words
In her native tongue to suffice,
But of course
When discussing particulars
Known only to her world
We had to use her language,
Which for example
Had hundreds of different words
For all the various kinds of light and heat,
Although none for weather
Since it never rained or even got cold,
There were certainly no words for night,
Blackness, stars, or for other worlds.

She continued,
"We have been together several years now,
And still I awake each morning eager to learn
Of new mysteries. Is there no end to knowledge?"

"Oh," I replied,
"Where I come from there is truly no end,
But one cannot possibly know everything,
So one ends up finding out things
Only as they are required.

"Oh the wonders I could have showed you on Earth:
The colors, the mountains, the forest, the meadows,
The scents, the tastes, the inventions.
I'm sorry that I don't have any books with me

Or even something so amazing as a mirror to show."

"A mirror?"

"Yes, you can see yourself in it."

"See myself? See another me? I cannot."

"Yes, it's like a reflection in the water—
Oh, I forgot—there is no standing water here,
And damn, I don't even have a shiny belt buckle
To use to show you the effect,
And all the glass in my spaceship is non reflective.
Anyway, yes, you could see yourself
Just as others see you."

*"From the outside of me?
I sort of understand
But I cannot quite imagine."*

"When mirrors first appeared on Earth
In the form of polished metal,
People thought them magical,
And even in modern times
One could watch with wonder
The amazement of babies or small kittens,
Who, though they both quickly got used to it,
Thought first that they'd seen
Another of their species."

"Kittens? Cat?"

"Small furry animals.
Domesticated—meaning tame or not wild."

"Animals? Wild?"

"They are other forms of life,
Some with four legs,"
I explained, ever so patiently,
For there were no animals on her planet.

It was in this way
That we often got nowhere fast with words,
But then all of a sudden progressed
With great leaps and bounds,

Especially with material ideas;
However, abstract concepts took longer,
And concepts like darkness
Were still pretty much incomprehensible to her.

"We had animals in the old days," she said;
"There are drawings on some of the cave walls
Of such as you speak.

"They are all gone now, like your Earth.
You seem so sad when you speak of Earth.
It must have been wonderful.
What do you miss the most?"

I thought for awhile,
Thinking of the scorched surface
Outside our cave.

"What I miss the most is not the darkness,
For I can simulate that here when I sleep,
And not love, for I surely have that now with you,
And not the cold, for I never liked it,
Nor life, for I am happy to have it here,
If nothing more;
But what I miss most,
If I had to say some particular thing,
Is the color green,
For green is a color
Which does not seem to exist here,
The hue that is the soothing and lush
Life-giving restful green of Earth.
It was said the be the sanest color,
Evoking serenity, as in your name."

"What is the color green?" she inquired.
"I know the blue sky, the golden suns,
The tan rocks, the brown leaves
And the brown vegetation,
The pink of your hidden parts,
The red of our blood and of your hair,
the orange flashes of fire,
The darker brown of trees that almost suggests
The strange black color that you speak of,
The gray shadows, and the yellow of phosphorous,
But I have never known there
To be a color called green.

What is green?"

"I wish I could show you, Serena,
But there is no green on your planet,
Not even a tint or a shade of it.

"On Earth,
The leaves and the vegetation are mostly green,
But here the same are all born brown,
Even in the shadows of caves.
Some people on Earth have green eyes even,
But alas, mine are brown,
And there is no other body part which is green.
Although nothing much else on Earth is green
But the vegetation, green has,
Even more than the blue of sky and ocean,
Come to be regarded as the sweetest color on Earth,
For it represents all that is living and supportive of life.

"It is very calming and serene, like you,
And therefore many people use it
As the color of their carpeting.

"Many of the other colors have drawbacks
Or more specific uses:
Red, for example means danger, blood,
But having red tablecloths in eating places
Makes people hungrier and so they order more food;
Pink is debilitating and so many of
The game playing sports teams have painted
The visitor's locker rooms in that hue;

"Blue is energizing and is
Often used in working places;
Yellow is bright and cheerful, the sun's color,
And is often used in cooking rooms called kitchens,
Although yellow can also mean caution, danger
Even, especially with black,
As on stinging insects called bees.

"Purple is used for mourning death
Or for the regal Kings and Queens, the rulers;
Our brown, like all around here,
Is actually the most popular non primary color
And is not therefore even in the spectrum,
For it is made up of red and yellow (orange) and black."

"But," she persisted, *"what is green like?*
If you can't tell me what it is,
Then maybe you can say what it is like,
Or perhaps you can say what it is not like."

"Either way that's hard to say,
For green is a unitary hue and also primary
And so being means that there is nothing like it,
No overlap; although if I had to say so,
I think green is more like blue than any other color,
But I only say that because green
Is a cool and soothing color like blue,
And not a fiery color like red, orange, or yellow.
But I should tell you that blue is certainly not green,
Nor is green blue. If I myself had not known green
Then I doubt I could have conceived of its existence."

"That is fine philosophy," she said,
"But it does not tell me what the the color green.
Have you any green clothes?"

"I do, or I did, but they're not with me—
Not even the slightest thread,
For I've already examined
All my clothes and space suits.
It's not that I don't like to wear the color green,
Although it is seldom worn on Earth,
Except on St. Patrick's Day.
I guess there was already so much green on Earth
That we all came to prefer more of a contrast.
And my spaceship, it's all metallic skin
And fiber optic conduit—
There's no green anywhere in it or on it."

"Patrick, I really wish that I could know green."

"Too bad there are no rainbows here or waterfalls."

"Rainbows?"

"Caused by water vapor falling from the sky,
Called rain, or fog or moisture
That divides white light into the colors
That join to form it."

"Waterfalls? Falling water?"

"There are none to be found here—
The rivers dried up long ago,
Leaving only those ruts in the ground
That I sometimes call canals.
I really miss green, though—
It is somehow a part of me,
But let's not give up on it so easily, Serena."

We ventured outside the cave
To begin a search for green,
There being only one sun up now,
Although it was still a scorching 120 degrees outside.

We lifted some rocks,
Finding various insects thereunder,
But they were either brown, gray, drab, or colorless.
I next peeled back some bark from a 'tree',
Hoping for some inner tint of mossy green,
But it was only tan.

I pulled up a plant and tore open a leaf,
But had no luck with that either—
Evidently chlorophyll played no role on this planet.

Next we cracked open a rock
But found not the fabled color.
This planet was indeed a greenless world.

"I guess green is not necessary for life here," she said.

"I guess not.
This reminds me of the time I read a book
Which did not use the letter "e",
The most frequent letter in our alphabet.
I didn't even notice it at first,
Although I had a vague feeling
That something wasn't quite right."

Even though I was now very tan,
I didn't dare stay out in the sun too long,
So we gave up our green search for today
And headed back to the cave.

"I wish I had some ice," I said.

"Ice?"

"Solid water—hard as a rock,
Nut so refreshingly cold."

"Make me some ice, Patrick."

"That I can never do in this climate.
It never gets cold enough."

"What is cold?"

"It's hard to explain if you've never felt it,
But it's sort of like how the underside of a rock
Does not burn your hand since it's cooler than the top;
Only "cold" means much much more cooler."

"I know not this thing called cold," she replied.

I looked fondly at Serena
As we walked back to our eternal cave.
Her skin was a very deep bronze, all over,
For she always went naked.
She had blonde hair,
But no other body hair,
Since her race had never known cold, I guess.
Her hands and feet were wide and leathery,
With six toes and six long slender fingers.

She had no real eye lids to speak of, or eyebrows,
But otherwise had all the other humanoid features
And anatomy; she was a human first cousin, perhaps.

She was completely vegetarian by necessity.
She had not a violent thought in her head,
Having, I suppose, no natural enemies to fear.

For dinner we gathered some brown leaves,
Which evidently contained
All the nutrients that we needed,
Since we were still healthy,
And then retired for the 'day'.

"What else do you miss from your Earth?"
She asked as we lay together.

"Well at night, for this is night to me now,
I miss the stars—those suns of other worlds
That I told you about, the stars that shine
Across the blackness of space,
For they are far away and thus appear very small."

"You mean that the space between the stars is black?
At least I think I know what black might be.
And the suns are not bright and blinding
Since they are so far away?"

"Yes but we will never see the stars here—
And how I miss them all,
That night sky full of lights.
I used to look up at it as a boy
And dream of going to the stars.
Then the transwarp drive was invented
And my dreams took wing."

"What do stars look like exactly, at night?"

"They are very small, just points of light really,
But they twinkle like jewels,
Such as diamonds and sapphires,
And some stars are even emerald green—
Like the close star companion of Sirius!"

"Jewels? Diamonds? Emerald? GREEN!"

"Jewels are stones that give off light
In colorful gleams and sparkles,
Like when you cover your eyes
After seeing a bright light,
Or like the gleaming sands outside."

"I see them in my mind.
Are there a lot of stars?"

"Trillions—so many that some areas
Of the night sky appear as cloudy white patches."

"I wish that I could see stars, Patrick."

"There are so many things that you'll never see!
If only I'd brought some photos of Earth with me!"

"Photos?"

"Yes, they are like permanent mirrors."

"I could see Earth and yet not be there?"

"Yes, and it would look about the same.
We even have three-dimensional holographic pictures."

"Oh, that there are such wonders!"

As I fell asleep in her embrace,
I had some dreams of Earth;
At least I could still go there in my sleep.
Oh how often had I taken Earth for granted,
Not appreciated it when it was there;
I even left it time after time
To go off into the cold and colorless void.
I tried to forget it, but I could not.

Well at least I was alive in this strange sort of Eden.
Anyway, life was life, and more and more
I realized that I didn't really need
Anything fancy from life,
Except love, of course.
Yes, love was enough—
And it is reason for all that we do.

Well, my spacecraft was still in working order,
But was there any place out there left for me to go?
Was there any life still out there?
Or had it all withered away by now
For lack of support?

There were plenty of
Borderline class-K planetoids around
Like this one, but unfortunately
None of them were anything like Earth
For a long ways in every direction.

Perhaps I could head in some chance direction,
Running out of fuel, of course,
But coast at a high speed for years;
No, it was too risky—
My ship was only of the intersolar type
And was not meant for distant star travel.

I could though, return to the main mining base,
An artificial world built on an asteroid;
But no good,
For it too was a world with an uncertain future,
A world even more sterile than Serena's planet.

No, my life was here now.
Anyway, all the greener worlds
Had nasty diseases and organisms
Against which I could never be immunized.

Yet another of those endless
Overly tropical days dawned,
But only in my mind,
For the sun never rose or set
Without another sun
Already in the sky before it.

No dawn, no dusk, no half light.
I had brown leaves for breakfast again!
Talk about the simple life!
I was becoming a modern day Thoreau.

*"Tell me again about the mysterious colors
Of black and green,"* she requested.

"Black is easier to tell of,
So let's start with it,"
I answered.
"Black is the absence of all color
And so it is the opposite of white,
Which amazingly is the sum of all colors,
Although white reveals not a one of them,
Except through a prism.

Since your life is based on phosphorous,
You see a dim yellow even as a background color
When you shield your eyes.
But when I close my eyes
I see a black background."

*"I can't think what would be there
If not for the yellow glow."*

"I'm afraid I'm not doing a good job of explaining."

"Try harder," she encouraged,

Then it hit me! "What a fool I've been," I said.
"I have an old-fashioned ink pen
Somewhere in my spaceship,
One that writes in black
On something old called paper!"

At once I retrieved it,
And the pen still worked
As I ran it along my skin.
"That is the color of night, my dear.
This is black. See how dark it is?"

"I see now," she said. *"It is as I thought,*
Being the limit reached by the removal of all light,
The color hinted at by shade and shadow,
The color just past brown, at least for me—
A lack of color really, like you said;
But I still do not understand what is green.
Do you have any green ink?"

"No, people don't usually
Write with green ink.
Red ink, yes."

I quickly ran outside
And went through my spaceship
With a fine-toothed comb.
The ship was all white and metal gray;
There was not a shade of green to be found
Anywhere on it or in it.
The seats were simulated leather
And all the electronic readouts were orange on blue.
All the supply kits were yellow
With the insignia of the mining company.
I had trouble even finding anything blue on the ship.

"We're a long way from Ireland,"
I said, exasperated.

"Ireland?"

"It's a country on Earth
Known for its forty shades of green."

"I wish that I could see Ireland, Patrick."

"None will ever see it again except in memory,
Although I came from there."

"Can you not make me the color green somehow?"

The question struck me dumb,
For it was really a very good question as asked.

"Wait a minute," I said.
Green is made from mixing blue and yellow—
But unfortunately I don't have any paints
Or such mixing materials,
Although I do have a bluish-black pen,
But of course not a yellow one,
For who would write with yellow ink."

"Worse than writing with green,"
She added, smiling, catching on.

"Yes, writing with yellow ink is silly
But yet there must be some way to produce green.
Serena, can you logically mix blue and yellow
In your mind and then imagine the result?
No, forget it, that doesn't make much sense,
For yellow and blue give no hint
Of the resultant green like, say,
The way yellow and red readily hint
Of the resultant orange."

We sat silently for while, stumped,
The heat growing stronger all the while.
She looked down at the ground, disappointed.
I too looked glum for a time. Neither of us spoke.

We saw nothing but yellow phosphorous
And the yellow sand gleaming even more golden
In the light of the twin suns—
There was yellow everywhere we looked—
Hot warm glowing yellow and more yellow
Until it had fully saturated the eyes and the mind.

Then I noted a flash of inspiration on her face.
She smiled and suddenly looked straight up

Into the bright blue sky, as none had ever dared to,
Then covered her eyes and screamed with delight.

"I see green, Patrick!," she cried.
"I see it. It's green!"

I quickly did the same as she, and yes,
The mixing of the blue sky
With the yellow afterimage
Of the phosphorous ground
Had produced a clear and vivid green.

She had at last
Beheld the verdant color of Earth.

THE MISSING LINK OF LIFE?

Now, how does one draw a clear line
Between organization and not?
When even does the night turn to day?

The most interesting and potent things,
From the evolution of the universe to life,
Exist at the blurred boundary
Between order and chaos,
Life perhaps emerging in tide pools—
The shifting edge between land and sea.

It is the fuzzy realm
Where things have to be
Orderly enough to take form,
But not so much frozen
That they cannot change.

THE FOUR ELEMENTS—AND DUST

From the fires of stars to those of cremation
We have breathed, flourished, and dissolved:
Life is ashes to ashes, stardust to stardust.

Of airy winds, vapors, and a soft earth
We rest at last under the spinning skies,
Those of Earth's sunny days and starry nights.

WILLING THE WILL THAT WILLS?
[Revelations: Austi: 4:5]

What is the "secret" of human behavior,
One that's really so much the saviour
That we may even keep it from ourselves
Rather than very far into it try to delve?

What is it that should be so confidential,
Classified, and undisclosed—its potential
Kept under wraps, so very contra;
Informally: hush-hush; formally: sub rosa?

Well it's a revelation of splendor,
One that's often good to surrender
But is also very well to remember.

Is the will free to will one's actions otherwise?

Can antecedent conditions be ignored?

Can the self be an unmoved mover?

No, but...

And what of those tendencies of evo's realm
That have been imprinted on one's genetic film—
Those of temperament, role preferences,
Emotions, responses,
And even one's most revered moral choices—
Those invoices from which one rejoices?

Well these are not choices
At all in and of any free will voices;
In essence, from the basis of one
And from all that one has become
From life's total behavioral reactions,
There are probabilities of actions—
Some patterns that are very likely
And some patterns highly unlikely.

Is free will a necessary fiction,
A kind of a religion?

No and yes if it's to provide an essential berth
For one's morality, meaning, and worth.

So then, with this "free will" become,
One might then succumb
To systematic deception
About one's causal connection
To that of nature,
A roadblock, a detour
That's neither possible,
Necessary, nor desirable.

The enemies to these "free will" motifs
Would be the mythical cultural beliefs
That explain behaviors and feelings
In terms of unknowable forces and beings.

But to protect one's moral virtues
Should one still believe oneself's purview
To be as an ultimately responsible agent, lo—
A self creation ex nihilo,
A God-like, miniature first cause who chooses
Without it being determined by one's own muses?

Well maybe but nay, really not, nil,
For there is no contra-causal free will.

What the good then
Of this fix we're in?

Such it is then that we can gain a measure of peace
Rather than the anger of resentment's crease
When someone does or says something bad,
Even those close relatives you once had.

For the civil-law-breakers
And all those ungiving takers
We'll no longer incarcerate
For punishment, being so irate at the jailbait,
But so that society will be protected
And that they might emerge corrected
From the swill of a prison mill,
Fulfilled with a new unfree will
That points more toward goodness
Or at least away from badness.

Thus the action
Of metaphysical justification

For a total retribution
Then greatly softens,
A relief from the stress so often,
For it's no longer induced
From the abuse produced.

Really?

Truly.

Indeed, we become less self-conscious,
More playful, less noxious, more gracious,
Less callow, and less likely to wallow
In the sorrow that is so hollow and shallow
In its excessive self-blame, pride,
Envy, or resentment—now all put aside.

Aren't we changing the will here as we go?

Yes but mostly no,
For the will must ever follow what we know.

Then we are learning—
The only hope for larger earnings
From the will's then wider yearnings!

Yes, we're overturning.

What if to learning we are averse?

What a curse! Might as well call the hearse.

So then, all in all, though a tempt,
It is that we humans are not exempt
From the laws of physics—as a preempt
Although we've been wired to make the attempt—
A seeming violation by nature
Of its own universal law and structure.

No it's not a violation I would call,
For science still did tell us all.
It's all part of the structure;
One can never cheat Mother Nature.

Hail, then, to the physic.

Well it's not so bad, is it?—
Although we can never will the will,
Its motives ever our intent to fulfill;
It is that we have no free will.

True, plus we can expand the will's horizoning
Through our broader learning's wisening.

Yes, learn today and by tomorrow, say,
The will may have a different sway.

I wouldn't want it any other way,
For then I wouldn't be me—my screenplay.

What other ways can we improve the play?

Well, we have patience and delay,
For we don't have to act right away.

Until a more creative solution appears?

Yes, from any frontier, Shakespeare.

Hear, hear!

DIPLOMACY

The best diplomat is one
Who can tell someone
To go to a place really warm,
Such that they feel that
They will even enjoy the trip.

**FOR THE SUPER
HEAVYWEIGHT CHAMPIONSHIP
OF THE UNIVERSE:**

GOD VS. SCIENCE

Round 1

In the Beginning...
God played an active role in the Universe
After creating it, each and every verse,
And especially the one upon the Earth...

Which is supposedly
Only a few thousand years old,
Or so it has been told.

God won this round, hands down,
For even those many science clowns
Who were there at the time
Thought that man was prime,
Being the special center of creation,
And that the sun and the stars in elation
Revolved around his nation...

And furthermore
That evil spirits caused physical ills,
Along with all of our mental ills,
As aggravated by life's frills—
Which were all called 'sins'
That somehow still came from within.

Even fun was one of sin's evil cousins,
For the Bible was made of old Jewish legends.

Thankfully those hundreds of odd Gods
Who had come to reign before GOD
Were crushed as by Jehovah trod.

However, about three centuries ago,
The realm of natural law was extended, so
The Supernatural Kingdom
Began to shrink away some,
Eventually vanishing from all of existence,
But we get ahead of our own persistence...

God made Adam fully formed, without a navel;
But now, an asterisk on page one of the
Philippine Catholic Bible says, "No",
Do not take it literally; it's just not so."

Round 2

God came out quick, still claiming the writ
That He guided the Earth safe through its orbit
Around the the centered sun in space, His Son,
For by now the Earth's motion around the sun
Was known to be true to nearly everyone.

Newton demolished this notion
With his laws of motion.

God thus no longer ruled Nature's course,
For the world was free to run its course.

From Isaac: Laws and Revelations:
There is a mote in space known as Earth,
A pale blue dot of fluff orbiting a hearth.
Due but to Newton's laws of motion there's none:
No Godly hand guiding it safe around the sun.

The vanishing had now really begun.
The heavens and the Earth were one.

Stars and galaxies went on and on puffing
And we became the center of nothing.

God was losing his definition in stone,
As his sworn traits disappeared one by one.

So He's retreated to higher ground, that is,
Outside of space, time, and all that exists.

Round 3

God so then claimed to appear to us
Only in spiritual thoughts and ideas, thus
Making Him responsible, as our Savior,
For the goodness of human behavior.

This metaphor was then found to be unnecessary
As the source of human character non-contrarily;

Yet some still clung to the life-line ropes
Of His intervention with their hopes,
Although some claimed that He
Did not involve Himself or be
In our daily operations and pleas.

So God no longer intercedes in causes,
Except in some nebulous cures and "becauses"
Like being safe from harm or curing what hears
But He never heals amputees, or appears.

For the latest is that He must stay hidden
Even if the "miracles" of His Son bidden
Were very much out in the open to see;
Better that no one know of Him clearly.

So there is "faith"—a blind trust in the unknown.
Believe it or be tortured—or has this too,
The Word of God, become inoperable?
Only the supernatural realm remains.

Round 4

God was still yet "seen" to intervene dear,
Saving lives here and there
In the natural world's reality,
But too, striking planes from the sky,
Ever adjusting and smoothing the operations
Of natural law—expressing His inscrutable purpose.

However, scientific knowledge, cosmology,
Fundamental physics, chemistry, biology,
Anthropology, and psychology were wont
To undermine religious views on every front.
God was losing His strength to be,
For science loomed large, quite ponderously.

Religious knowledge without proof,
That which professes absolute truth,
Now fails and fades, an impossibility,
While science, which professes fallibility,
Succeeds and grows stronger daily.

There were still those strange myths...

Why is the Old Testament out of the pew
In many churches, in favor of the new?

Was it divine revelation or not?
Do God's fits not become a good role model?

Round 5

With God in full retreat it was yet thought
That at least He had instilled or wrought
A spiritual essence in us willed, whole,
That which was called the "soul".

What vanity to claim such full self-importance!
To demand so much from the universe...

That one would claim an angelic vapor that
Drives a living being, provides character,
Morality, and consciousness, on top of
A burdensome, fragile, and expensive
Organ such as a brain not needing to be used?

Science collapsed the idea of the elan vital
When the synthesis of substance began.

Life's chemistry was of chemicals!

Yet, it was still said that God made all the kinds,
Albeit strangely full of the problematic signs
Of such an unintelligent design,
For how else could it all have been consigned?

Darwin told us how natural selection
Explained the mysteries of evolution
And of the variety of life covering creation...

Extending from animals to us, a continuum,
Now even seen to go back to a bacterium.

We were no longer special at all, as such,
Differing from chimps by not very much.

The discovery of genetics later on
Collaborated it all in our genome.

So because of evolution's record written
God's Bible was no longer seen as written
In plain text for the common man,
But is now open to symbolics and interpretation.

Thus now He just is, the same as the universe,
And yet this would be a kind of curse,
For this state would be quite restrictive...

Not to mention the mere tautology
Of a universe, a cosmos, and an Entity
Being one and the same pose,
Such as a rose is a rose is a rose.

Since the above Cannot be,
He's now become but a Deity,
Leaving us all on our own,
Our own life to own,
The same as the nonreliance
That is seen by science;
Now we're fully sentient,
But a planned, random accident!

Aye, the truth of what now we are is:
We're not made direct by a Wiz to take a quiz
But as mammal, organic, of speciation—
One passing narcissism and self-adulation,
Onto the bio-electro-chemical organism
Evolved upon a planet near a star, risen
Of and in the long and winding mindless way
Of slow time, dust, and selection by death
That sifts the best from the rest: evolution's breath.

Round 6

More devastating blows landed, raw,
Einstein's theories extending Newton's laws
To the very large universal scales, with trust,
While quantum mechanics brought us next
To the reach of the very tiniest of objects,
There being no place left for us as subjects.

God was nowhere to be seen,
Having vacated the arena.

Yes, science has found that the universe
Operates just as it would without Him—
That evil spirits don't lead to bad health,
That brain imbalances can lead to sins.

Devil, Hell, the Bible, intercession, etc.,
Are all gone now; he is undefinable—
Protected from the knowing—safe, away:
Yet claimable as the unseeable unknowable!

Round 7

Confirmations were everywhere hatched,
Since scientific laws must ever match
And predict the facts of what it mimics,
For example, of the quantum mechanic.

Although QM's basis seems counterintuitive,
It always works out just perfectly,
For we employ and depend on it in every way
On tech products based on it, every day.

Science ever goes on to astronomical heights.

The first supernova since 1572
Appeared in some small galaxies nearby, a few,
Called the Magellanic Clouds, too...

Though its radiation began a while back,
We saw it alight upon us in the 'now'—
The immerse quantities of energy
Of a mighty star-stuff maelstrom.

A Chilean astronomical technician, some bloke,
Stepped outside, perhaps to have a smoke,
And being observant spotted its yoke!

Ah he, a mere human standing around
Out under the dark starry sky, aground,
Detected it upon this lucky time,
For the large telescopes only take in the shine
Of the sky in small sections at a time.

He went in and told of such unexpected,
That a large burst of never-detected
Neutrinos was now to be expected.

The astrophysicists called their colleagues,
C'mon, you all, answer, please,
Those deep beneath the Earth's surface,
In the United States, Japan, and Europe,

And then said, "Look in your tanks, in revelry;
You have already made a great discovery."

They were right on the dime, this time;
Each of the observatories had detected the signs
Of a few tens of neutrinos at about the same time.

Consider the magnitude of this achievement,
For they had tested all of what physics meant!

They had predicted the events that go
On in a star's death throes
By using theories from nearly every part of physics:
Special and general relativity, quantum mechanics,
Fluid mechanics, thermodynamics, nuclear physics,
Atomic physics, and elementary particles.

If any of these theories had in error flailed,
The prediction of the neutrinos would have failed.
Thus the laws of nature that are known to us
On Earth everyday must have the same thrust
Hundreds of thousands of light years away;

And also the same back in the day
When that star had exploded so,
Hundreds of thousands of years ago.

God had been pushed completely out of the ring;
Now there were no more praises of Him to sing.

There were no immutable forms made,
As is, that never change, as "bade",
Since there was no one miracle of life
Leaping into any living form, but rife
With all of natural selection's strife.

Slightly dead chemicals
Became definitely alive chemicals,
Metabolizing into many particulars,
This being not so spectacular.

We even have evidence of ancient algae
From 3.5 billion years ago, in a sea,
When liquid water was available and free.

It still took more than two billion years
For more complicated life to appear.
This was no Garden of Eden.

God's become aloof; he's begun to dissolve;
He let the design gradually evolve
Over thirteen billion years into man's plot,
The endless universe a mere backdrop.

He is the Intelligent Designer that
Is deducible from not understanding design,
But wait, He is of infinite design—
So now we know that something had to make Him!

Round 8

The Knockout.

The three-degree blackbody radiation was found,
The CMBR; it comes to us from all around.
Nonuniformities in the radiation were found at last,
Those that formed the galaxies of the past.

The QM realm has been proved of late
To be a fundamentally fuzzy state,
Virtual plus and minus states
Popping out at any old rate;

That is, there are no real causes,
For there are no hidden "becauses".

This realm is not quite a Nothing,
But a near 'nothing',
Or an infinite regress of something.

Virtual particles may take the helm

Or cancel back into the QM realm.

If "Nothingness" were exactly zero, not fizzy,
Then this 'vacuum' would not be vague and fuzzy.
Thus an absolute Nothing cannot exist to be,
For its very definition means that it cannot be,
As then it could not even be there at all in reality.

So there is ever the quantum jitter;
There was only this 'possibility' forever.

Oftentimes the QM "virtual" particles magically
Spring into existence and vanish quickly,
Although they can interact and remain, really.

If not they have to vanish so quickly
That we cannot account for their reality.

If we could see them then the QM possibility
Would not be the vacuum fuzzy energy;
But if they were not there as something
The vacuum would be exactly Nothing...

And so this certain school
Would violate the vague and fuzzy rule.

None of these happenings are invisibly lame,
Such as those of the supernatural claims,
For the fuzzy 'nothing' has many effects
That we can compute and detect.

So is there is no cause, no purpose, unthunked!
Does this make us go into a deep dark funk?

No, for it is our glory that we are free to be,
The making of life being our own responsibility.

Now God was dead, gone, having counted out,
Having never been, whether within or without.

The eternal, causeless ground-state
Could have never had any "create",
For there could be nothing prior
Such as that which is known as a Creator.

Terrorists still go to war in his name;

It's all going astray—this notion fails;
If I knew where the Great Designer stays,
I'd question his mysterious(insane) ways.

What then is left of this vanishing Phantom?
More features than I've listed have fallen—
The Extraordinary Superstition's kiss
Remains as but a shadow of a wish.

PROOF FOR '40'
BEING THE BASIS OF ALL

Abraham Lincoln,
Biblically named,

Wrote

"Four score and seven years ago"
As part of the Gettysburg address
On the back of an envelope
(Which was the wrong side)
On the way to the battlefield.

Note that "seven" is almost "eight"
Which is 2 x 4
And that 8 x 5 = 40.

Also, the Holy Trinity
Has been expanded to include Darwin.

4 x TenCommmandmens = 40.

76 trombones + 4 = 80 = 2 x 40.

QED

THE TIME CAPSULE

Since one million years had just passed by,
They of the future prepared to open nigh
The absolutely sealed container's prize
Of the capsule made so carefully that it did survive
Without damage, being totally impregnable
To any outside influence imaginable.

They expected to see perhaps some old relic,
But certainly nothing alive that could tell of it,
For it would be hard to imagine even then
That some organism could keep on living its ken
Over its course onto a million years later,
Sealed inside this tight container,
Unable even to exchange energy's spark,
This metabolism being the hallmark
Of life and all that quacked or quarked...

And so they did not at all expect something
In there that would be flapping its wings,
Gasping for air, or anything at all of life's song,
It wondering what had taken so long.
Well they were right and they were wrong,
For in the time capsule planted ago so long,
Several things had with it come along...

One was a plaque, of numbers low and high,
It containing some primes and pi;
Another, some essays of the future—
Some, like Austin's, quite mature,
Along with Darwin's book, maps curled,
And many other items of the world
From those times when the oceans swirled;

But the last one, perhaps not intended,
Was a microbe—an extremophile,
Sitting there quite contented all the while!

Well they soon laughed, loud and long,
For they were in between right and wrong
As to what could survive from so long ago,
For it was really walking mighty slow!

Stunned, twice they had to look;
It had crawled out of Darwin's book.

MAGICAL HAPPENINGS

What secrets of life and death
Lay buried in the sands' depth?

What inaccessible truths
Protect their own proofs?

General Rascal lit up a cigar,
And the stories unfolded far,
In the haze of a pipe dream...

"Do tell us what under lid
Was in that Great Pyramid."

"There were 4000 years-old iron weapons
That did not rust, from the famed in wars,
Looking as new as the day they were forged.

"I folded glass that bent without breaking;
I beheld visions of no Earthly making.

"I drank from a vase of flower stems
That poured water without end,
And filled an entire tub from it
Bathing away all my dirt and dust.

"A compass needle went around
And never stopped, I found.

"I ate a cake but I still had it.
Yes, I had the cake and Edith too.

"I saw the starry skies and all
Through solid rock walls.

"I entered a room that had no doors
Or windows but just floors.

"There was light within the room
But no flame or openings to those tombs.

"I looked into a grain of sand
And saw eternity and neverland."

He paused, recalling.

"Outside, I saw the Sphinx.
Its glance was fixed on something else.
It was the glance of a being
Who thinks in centuries and millenniums.
I did not exist and could not exist for it,
For it was the face of eternity."

ENERGY MATTERS

Perhaps matter is more than just equivalent
To energy, if it were transformed,
And more than equal to it, more or less—
It may be that matter IS energy.

Perhaps energy is projected by
Information or is a part of our
Reality illusion, as well, but,
We can't stand on turtles all the way down!

Only four percent of the Universe
Consists of the matter we know and love,
The remainder being hidden from us,
So, perhaps, this is what's really in charge.

The basis of the Universe was forever here,
For nothing can make itself from nothing at all;
Such, a state of nothing could never be, for there IS
Something—something that consciousness interprets.

Mind and matter are made of the same stuff,
That "substance" made only out of itself.

Mind experiences the present moment;
Matter records the present from the mind;

That is, Present Mind, Past Matter, combine
The frames of Space and Time into the film
That lives and plays in us as Consciousness,
Mind taking Space and Matter doing Time.

Well, how can I save the Soul—Consciousness?
It may create potentials, quantum-like,
That give rise to the Reality of
The Mind and Body—so, use it wisely.

THE SIMPLE BASIS OF BEING

As for forces, which are just a prelude here,
We note that two of them are transitional,
The Electric and the Magnetic,
Each giving rise to the other,

And that two others are oppositional,
The Weak and the Strong,
The Weak promoting changeability,
The Strong promoting stability.
Gravity is then left as a blend of all.

(Strong vs. Weak) [Gravity] (Electro <—> Magnetic)

So, would oppositional and transitional pairs
Work for our human being as well?

For us humans, all is of the
Movement of *Appearances,*
These *Movements* giving rise
To notions of time...

Past into Future,
Or the Then to When through the Now:
Transitional in only one direction;
While *Appearances* beget notions of
Matter lumps, in a place of Space...

Matter and Space, or the What and Where
Are a kind of an opposition in that
The knots of matter are separate
From the gaps of space in between;
Or in short all seems to be the
Movement through time/distance
Of Matter in Space.

(Matter vs. Space) [Being] (Past —> Future)

We will see that our being is composed
From these simple notions begun,

For *movement* grants time—
The Then and the When
Of the Past and the Future,
Via change;

While Matter is the What,
And Space is the Where,
Via 'clumps'.

The blend of all these would be
The spirit of life.

These fields then further combine:
The What/Matter + When/Future field
Becoming the Progression
Of matter into the future,

And the What/Matter + Then/Past field
Being the History
Of the matter past—what has occurred.

The When/Future of Where/Space field
Makes for Wishes, hopes and dreams
In the future place of space;

The Then/Past + Where/Space field—
Begets Remembrance of memories
In the past space.

The emergent fields then further combine:
Learning becomes of Remembrance and History;

A Change of Outlook becomes
Of Remembrance and Wishes;

A Change in Structure is Progress from History;
And *Vision* is of Wishes and Progress.

At the next higher stage,
Being **Creative** is brought forth
From *Learning* combined with a *Change in Structure;*

Direction results from *Learning*
And a *Change of Outlook;*

Growth is the *Vision* for a *Change of Outlook;*

Planning is the *Vision* for a *Change in Structure.*

Finally, **Creating**, **Direction**, **Growth**, and **Planning**

Compose one's being—The Who.

(summary):

Nothing (Why) —> Possibility (How)

!
V
{ [Space(Where) <— Appearances —> Matter(What)]

SEGNO (SIGN) # 0

"There is the 'vacuum'", replied the other,
"A base state, one pervading all of space,
There being no signposts within it,
Or anywhere, since it is of no direction.

"We must regard it the stuff of which things are made;
For just as all living creatures inhale the air,
So do all the real natures inhale the vacuum."

"This intimation is the mark of manifestation,
A demonstration that's the token of the evidence;
The aetheric and heavenly sign of things to become,
Both the portent and the omen of so much possibility.

"It is both the warning and the present notice,
Presaging both the promise and the threat.
Aft this sign, that the vacuum 'indirects',
Then the real gestures ever beckon;
They of an the unsignal faint,
The wave and gesticulation of you.

"We read the noise of the quantum theater—no marquee;
All is daubed without symbols, to mark no cipher, bare,
No letters, characters, figures, or hieroglyphs there,
No ideogram of the rune of order,
No emblem of the Divine."

GRACIOUSLY WELCOMING
LADY LUCK BECOMING

He believed that luck would never fail—
So he ran like the wind through the jungle,
Surely knowing. He'd what he'd come for,
Now hopeful to find the help at the shore.

The relentless ones were not far behind,
That ill-fated menace of the bad kind.

Miss Fortune laughed, and said,
"No road could be too hard to tread,
For we are fearless. To those, a boon—
For they ever seize the Opportune."

"I see you, Fairest Happening."

Just past a sharp turn, in the trees
He suddenly dropped to his knees
And fired into his pursuers mean
As they came upon the scene,
Using all his ammo but for one round,
Then hurried on with nary a sound.

"I am wide aware," Miss Karma,
"Of this continuing Dharma—
That chance shines as my sun,
For she in turn, happens on everyone."

"Oh say it is your lot, my friend and lover,"
She answered back, granting him cover.

Listening, he could hear ever more troops
Rushing through the night in groups,
About a half-mile back around the loops.

"I gratefully welcome thee,
Miss Lady Luck of Dice,
Though I may pay a late fee
For my pick up so precise."

Ms. Destiny Serendipity smiled, saying,
"The game is on; we are playing.
Let joy and innocence prevail;
Believe that luck will never fail."

He moved on, ever faster, cheating death,
A third wind becoming of her vaporous breath,
It blowing this DIA operative onward
To the shore ever toward.

He could hear the whirling chopper,
But now receding was its Doppler,
He thus grieving of its leaving.

"Am I much too late—still too far?
Shall I curse you all, destined stars?"

"No," said lovely dear Twist of Fate,
"For you have one bullet left for chance,
Not to use to sleep or dream perchance."

But the chopper was rising high,
Well into the star-crossed sky.

"Shall to self I take this bullet
Now that the bus has left?"

"Oh, no," Miss Lucky Break encouraged,
"Do not be at all discouraged,
For you know it shall not be so
And what with it you now must do."

"Yes, perhaps it shall be so in some plight
Coinciding in a most kempt and hapful night."

He smiled and then knelt to ground,
And sent his last bright tracer round
Just ahead of the copter now departing,
His minor wounds yet sorely smarting.

"I bless you with all my lucky charms,
My good and well-fated man of arms."

The door-gunner noted the red tracer
And whence it came of the river vapors.
"Captain, turn back and take a look;
He awaits a fortuitous accidental fluke."

"I am an uncursed, non-jinxed agent man.
Let my joyous innocence prevail again."

He jumped into the rescue's hovering haven,
Directing the door-gunner's firings, wavin'.

"Fare thee well, my nightly knight,"
Dame Fortune wished upon his sight.
"You recognized me even in the dark."

"Oh my Angel, you lovely lark,
I might have known it was your spark
That would ever see me through."

NO TIME MACHINES

Now, tongue in cheek,
I'll tell why there are no time machines.

It isn't that no one ever came back
From the future to see us,
Although that is still a good reason
For no time machines being possible.

Nor is it that there can't be
A future going on somewhere ahead of time,
As that's a fine one, too.

It is that women prevented time machines
From being invented;
For every time a man said,
"Honey, I'm going out to the garage
To work on my time machine"
The woman in his life would reply
"That's impossible, dear.
Stop wasting your time;
There is housework to be done
And grass to be cut."

The man would still sneak out
To try to work on his time machine,
But the woman would find him
And once again say something like,
"That's impossible, you nut head.
Get in here and do something useful!"

And that's why there are no time machines!

HOW THE ALLIES WON WORLD WAR II

Warner Heisenberg, the head of
The German Nuclear Weapons Effort,
Was full of the uncertainty
That he had discovered in physics.

Heisenberg was entangled with his old mentor,
The Danish physicist Neils Bohr,
They being old friends, like father and son.

They were also supposed to be enemies,
For Germany now occupied Denmark.

Together they had created a physics
Of deep truth and beauty,
For beauty was the expression of truth.

They also made possible the physics
To destroy large cities, even the entire world.

In 1941, Heisenberg went to see Bohr,
The 'Father of Quantum Mechanics',
In Copenhagen, Denmark,
But we don't know what they discussed,
Yet Germany failed to complete its work
To build an atomic bomb.

Did Heisenberg deliberately withhold
Information from the Nazis?
Did this consummate mathematician
Neglect to perform an obvious calculation?

Did he, with Bohr, form a complimentary pair,
Joining their views of the political position
Versus its velocity to form a complete picture of reality?

Did a man's heart turn the tide of war?

The Drawing

On september 9th, 1943,
Neil's Bohr walked to a meeting place
Near the water and crawled

In complete darkness to a beach,
For the Gestapo in Copenhagen
Were about to arrest him.

He secretly crossed the Oresund to Sweden
And remained there until October 6th,
Wherefrom the British flew him to Scotland.

That evening, Sir John Anderson
Gave Bohr a briefing on just how far
The Anglo-American Atomic Bomb Program
Had progressed.

Also, Fermi's reactor had begun operating
On December 2, 1942.

Bohr was shocked, for he knew that only
The very rare isotope uranium 235 had fissioned
In the German Hahn-Strassman experiments.

This was fully two years after
Bohr had met with Heisenberg in occupied Denmark.

What had the Germans done during this time?

No wonder Bohr was alarmed.

And yet Bohr somehow
Had a drawing of the German nuclear reactor,
Which at first he thought might be the weapon itself.

All knew that plutonium,
Which does not exist naturally,
Could be chemically separated
From its uranium matrix
After bombarding a reactor's
Uranium fuel rods with neutrons.

The critical mass was not in tons but in pounds,
Thus prompting the allied effort,
Not so much Einstein's letter to Roosevelt.

Bohr went to work at Los Alamos
Where Oppenheimer was orchestrating
The impossible from 1943-1945.

On New Year's eve of 1943,
Scientists looked at Bohr's drawing
Of Heisenberg's nuclear reactor
In Oppenheimer's office.

Within two days, General Groves,
The military commander of the project,
Received a document beginning with

"The proposed pile [reactor] consists of
Uranium sheets immersed into heavy water."

And ended with

"The arrangement [the drawing] suggested to you
By Bohr would be a quite useless military weapon."

By late 1943, nearly everyone
In the German nuclear program,
With the exception of Heisenberg,
Had become convinced that uranium plates
Were inferior to a design using rods or cubes,
For the most efficient design
Involves separated lumps of uranium
Embedded in a lattice within the 'moderator';
But the worst possible solution
Is placing uranium in sheets or layers.

The role of the 'moderator'
Is to slow down the fissioned neutrons,
With only heavy water or carbon
Seemingly being feasible.

The Germans had chosen heavy water,
Its separation from ordinary water
An expensive and difficult process,
Since carbon graphite is rendered useless
By an impurity of as little
As one part boron in 500,000.

At Los Alamos, Leo Szilard was a fanatic
About the purity of the graphite,
And since it was readily available
They decided to use it for carbon.

The dragon's breath was to be unleashed.

None of the German reactors ever even operated.

Where did Bohr's drawing come from,
For it had "Made in Germany" written all over it?

It could have only come from Heisenberg.

The Further Whims of Fate

In 1935, Fermi had almost discovered fission
Three years earlier than Hahn-Strassman,
But in order to shield the detectors
From unwanted radiation
From the slow-neutron process
He had covered the uranium target
With aluminum foil.

This prevented him from seeing
The very energetic pulses
From the uranium fission that was taking place.

Thus the race to build an atomic bomb
Might well have started in 1935 rather than 1939.

If so, World War II could have been nuclear
From the beginning or even have become a cold war—
All of this not happening because of some aluminum foil!

THE EVER VICTORIOUS

Over Man came the Triumph of Love
But Chastity gave it quite a shove;

However, Death then all conquered,
But this was not the final word,
For Time happily reigned over all,
Or so it thought—as its thrall,

But Divinity vanquished its trend;
Yet still this was not the end...

For as ever the basis was left to sting,
Since Nothing overwhelms everything.

ELFIN LEGENDS

I love that time just after the gloaming,
With the last of the glowflies still roaming,
The night airs full of amorous promise,
And the quietus that lets love's dreams rise.

Here our ships are over-brimmed with visions
So clearly seen that some must roll on in,
For the prismatic arch of the sun bow
Is anchored twice to the real world below.

I dip myself in, as the cup to fill
From the stream of consciousness my will
That is beyond the plain reality,
As waking from it all the more to be.

The gossamer mist snared me, barely felt,
A fairy cloth of prehistoric weave that yet
Haunted every stream, meadow, and wood
In the Land of Youth, where timeless beings live.

For as I'd sensed the cloud of lilac fragrance
From a mountainous bush, that passing mist
Awakened olden creature things in me
That sympathized on an ancient frequency.

To life's forgotten tides and swells I yielded,
And thus was allowed into their spaceless world—
Through a small opening that tunneled at first,
Then funneled into the expanse of Fairylande.

Trumpet flowers had announced my coming,
My ticket being the poems that I'd written
On the lore and legends of the flowers—of Eve
And elves bringing forth all that bloomed and grew.

The Lande appeared at first much like my own,
But I saw colors that I'd never seen,
That were neither blue nor green nor in between,
And further they shone in some strange direction.

So this tale I give you if I can return to tell it,
Of shadow worlds within Earth's dominion,
Preternatural places transcending time and space—
The enchanted Faery World, Earth's missing link.

Of man and angel, one yet neither, they came,
To dwell forever in shadow worlds, between
Form and substance, they, all elfin creatures
And all who float or fly as came from Paradise.

Yet neither here nor there, though everywhere,
They're the fairy host, nurslings of eternity
And of all things everlasting, like Amaranth,
And of all things Heavenly, with love and dreams.

Alive only at life's Heavenly cusp,
They appear but in half-light dawn or dusk,
Seen usually by some quick sideways glance,
Or through some autumnal haze perchance.

Fays live in a moon-blue star-mist, out of space
And out of time, rising each morn as vapors,
Wavering here-there in rainbow colors,
Being the light and life of all leaves and flowers.

Midwives of bloom, wizards of natural miracles,
Painters of green, and guardians of buds,
They, the keepers of what moves all things,
Are with flowers Heaven's smile upon the Earth.

Born of kisses, fays are life's spirit-soul,
So much felt as to be oft seen and heard.
Their musical wings play songs so intense
That they fall as perfume upon the sense.

Fairy tinklings are sensed as drowsy fumes—
Incense lifting one on wings of fairy sighs,
The tide that turns us—as seen in the wake
Of leaves rising in their swell on windless nights.

Other mortals I saw too while passing through,
Then knew that Elflande overlapped our own;
They were not asleep but frozen in motion,
Awake yet unmoving in their instant of time.

Such when each moment passed unto the next
These mortal beings passed too, wondering what
Might have been seen: phantasms in the mind
That fell between the frames of their living film.

All things I felt continuously now,
As in humankind I knew only in
Rare moments of ecstasy when melded
Happenings had lifted me heavenward.

Magical things I saw, that only appear
On Earth when one's eyes close but for a second:
Wingèd ladies, and flowered butterflies,
Whose prints are pressed as dust upon the pansies.

The birds were of a species never known
And seemed to share a special closeness
With their elven brethren, faery sisterhood—
Which I knew and felt and saw as kinship.

I heard woodlands that once only whispered,
Meadows where there was once but a murmur,
And grasslands unhushed, full of wondrous sounds—
The music from beyond the human range.

My senses were heightened: touch went deeper;
My eyes saw colors beyond the spectrum;
I reached into living things, knowing them;
All the odours called, mixed with emotion.

A flush of youth shot through me, as the chain
Of light from angel to faerie added my link,
And my eyes were sparks of bright burning fire,
Sense extended in a new dimension.

I sprouted wings, and flew like a bumblebee,
And fell in love with a lovely wingèd flower
That had come to life, the vision of fantasy,
Her elfin eyes beckoning me toward ecstasy.

Summer follows us around, elfin queen
Of hearts, as we lay snug in winter green,
Your glowing pixie crown lighting the scene,
Your curves spooning, your ears pointing away.

As fays we made love in the air, hovering—
Evanescent visions of disembodied happiness,
The magic link in the chain of things, connecting
Man to God by angel and star to all that we are.

Satisfied, fulfilled, yet desiring more,
We returned to our cabin, loving deep
Into sleep, as blackness fell all around,
But for the starry memories that glowed.

Although I can stay no more than a year away
Or lose mortal form, this place will be my home,
So here I'll return, the seasons going round,
Where I'll continue to expand this poem.

Although back to deliver these words, I already
Miss the sylvan solitude and the crystal pools
Of the enchanted worlds 'tween Heaven and Earth,
Where the wanderers of light call me home.

So now, live your life and dream a dream through
Dale, meadow, and field, in grove and greensward,
Across love's pure stream, in shimmering sheens
Of the dells of Elflande; it takes but a wish.

THE IRRATIONAL

Indefiniteness didn't sit well with Pythagorus,
Ever concerned with the perfection of numbers.

You can divide the circumference of a circle
With its radius but you cannot write the result
As a fraction or a ratio, for pi just keeps on going.

So, he pledged his disciples to secrecy,
For the ancient Pythagoreans
Had developed an entire religion
Based on the rationality of numbers.

Yet, one rebel vowed to let the word out;
However, he mysteriously drowned at sea.

THE VAULT OF EVERYTHING

A spirit led us onward, who knows how,
Toward the Library of Babel,
Which contains all the possible books
That could ever be written,

Including, for example,
Better and worse Shakespeare plays, brand new plays,
Books with only one word of difference among them,
Everyone's life story (even the parts not lived yet),

The Secrets of the Universe,
The true Theory of Everything,
A lot of gibberish, and so on, as we can't imagine.
In fact, I found this story in there,
In a short story book of mine-to-be,
So I just copied it to here. (Yes, it said that too.)

We passed through jungles and over mountains.
The spirits led; we followed.
We even floated above the clouds
And then came back to earth.

A clear night sky of infinite possibility
Showered us with photons,
Lighting our way to the fountain of all knowledge.

"True enlightenment awaits me there,"
I offered to the guiding spirit.

"Don't be so sure,
Although you might chance upon it,
For the deep truths of enlightenment
Are as needles surrounded and consumed
By the near infinities of the haystacks
Of deception and confusion,
For, remember, Everything exists in this library."

"It must be a massive building," I remarked.

"Well, yes, but it's bigger on the inside than on the outside;
Otherwise, it would have been larger than the universe."

"Bigger on the inside? How?"

"Well, you'll see, but I'm not sure how;
Maybe through some dimensional extensions,
Or perhaps it's constructed digitally
And expands as you move about, somehow,
To conserve space; but even with compression
It's still hundreds of miles wide in every direction—
On the inside."

"What is Everything, in principle?"

"It's every arrangement possible,
Given whatever constraints there are, if any.
Of course not all paths may be stable,
Sensible, or last very long."

"That's a lot
Why do we live on this particular path
That our Universe has taken?"

"Who the heck knows!"

"What about making the forms
Of substance of a Universe?"

"Well, in the case of the emission
Of the secondary substances, let's say,
It's every one of the 'alphabets'
That can be conceived by
The Timeless-Formless-Motionless,
Plus all of its resultant workable combinations
And interactions of substance,
For this Babel library,
It is every possible arrangement
Of words in every language,
With punctuation, too, naturally."

"Hey, here it is. I can't wait!"

We looked in, picked a hallway,
And headed towards the stacks.

We saw piles and stacks of books
In every direction, even up and down,
Stretching toward infinity.

"Where's the card catalog?"

"There can't be any, for many titles and descriptions
Of similar books are too long to differentiate.
Think of the books themselves as the card catalog."

"How's the library organized?"

"It can't be. It would take forever."

"Who runs it?"

"Borges is the lone librarian
But he's somewhere in the back
And hasn't been seen for decades."

"OK, I'll pick some at random."

(Hours pass)

"Anything?"

"No, mostly mumbo-jumbo, but I found one on a table
That someone must have treasured."

"Oh, yes, he spent his entire lifetime here.
t's Plato's 'Beyond Metaphysics'."

"Wow! That's been lost for thousands of years.
But is it the true version?"

"Who knows."

"This library contains
No information whatsoever!"

"True, but there's another library next door
That also claims to have Everything."

"You mean that little 'hut';
No, wait—I get it—
The library next door is empty."

"Yes, for the All sums to the None.
Both everything and nothing
Have an information content of zero."

"Wait, I found two more good ones
In the stack right near the entrance."

*"One is by you and one is by your friend, Rascal.
You put those there in the first stack
So someone would easily find them
And read them, even though they exist again
Somewhere else in the library."*

"Yes, and I'm even going to let them
Stick out a little on the shelf."

...

Epilog

...In another chilling Borges's story
I read the actual book that he refers to,
The one whose infinite pages are ever-changing,
For that's how books appear to me in night dreams.

Sometimes there are even digits occurring
In the middle of words,
Plus if I look away and then back
Then the contents of the page have changed.

One time, when the page stabilized
To quite understandable words,
I realized I was reading something very profound.
In fact, it was the Ultimate Answer:
Sound, letter, phoneme, word, phrase, sentence, (uni)verse!

I dared not look away
Nor try to copy it with a dream pencil,
But instead tore out the page
And crumbled it into my hand,
Then forced myself awake.

When I awoke I had the page in my hand, and it said:
*This page intentionally left blank,
Except for the above,
And the above, etc.*

BECOMING

We humans mirror and recapitulate
All of evolution while growing in our mother's womb,
Racing through the stages in which life evolved.

During this nine months and even beyond that
We move from mindlessness to shadowy awareness
To consciousness of the world around us
Onto consciousness of the self
And then even to becoming conscious
Of consciousness itself.

For the first two and one-half years of life
The inexplicable holistic world
Is experienced less and less holistically
As the child discovers the
Bounds of discrete objects.

The holistic right brain remains of course
For us to take in the overall view,
While the logical left brain is also there to recognize
The detailed relationships between objects.

As such, so goes the universe,
Since we are formed in its image.

So then this gives us a clue
To the nature of the universe.

Seeing that the brain is
Divided into two hemispheres,
Each with their own
Characteristic mode of thought,
Which can communicate with each other,
Means that we are looking very deeply
Into the way that reality itself is constructed.

These two complimentary aspects
To the cosmos are thus absolutely essential,
One being of the whole:
The apparently indivisible,
Continuous fluid entity
Although discrete at unnoticeable levels,
The other being the interrelationships of the parts.

Each interpretation cannot appear
At exactly the same time,
But the Yin ever gives way to Yang
And ever then back to Yin, and so on,
The rounded life of the mind
Thus continuing to fully roll
As the cycle of this symmetry
Turns and returns;
If not, one either gets totally lost
In the details or prematurely halts
After but an apparent whole.

The holistic right brain mode is unfocused,
As we see in some people
Who are unconcerned with details,
It always building the scene in parallel
To form a single entity;
Whereas, the focused left side of brain
Isolates a target of interest and tracks it
And its derivatives sequentially and serially.

Yet the two sides of the overall brain
Are connected to each other
And so the speed of the juggling act
Can meld them together
Into a complete balance like that
Portrayed by the revolving Yin-Yang symbol,
Each ever receding and giving rise to the other
Such does the universe go both ways too,
Its separate parts implicated
With everything else in the whole.

During conscious observation
The 'hereness' and 'nowness'
Of reality crystalizes and remains,
We establishing what that reality is to some extent.

We define and refine the nature of reality
That leads to the mind's outlook.

Counterintuitive? Cyclical?

Yes but it is the universe in dialog with itself;
The wave functions and yet the function waves.

The universe supplies the means of its own creation,
Its possibilities supplying the avenues
And the probability and workability
That carve out the paths leading to success.

So here we are, then and now,
The rains of change falling everywhere,
The streams being carved out,
The water rising back up to the sky,
The rain then falling everywhere,
The streams recarving and meandering
Toward more meaning and so on.

HOPES

Hopes flitter and flutter like butterflies—
Whose forms show there can be a second life,
Although still one chained to time's mortality.
Do we fly through time on a two-way street?

Not wishing to face an end to ourselves,
We can hope that whatever brought us here
Will someday, somehow, take us home, somewhere.
We should look at how we're thrust into life.

We're constructed from various star stuff,
Through life's history recorded in strands
Of DNA both recent and older,
All the parts joining, to work as a whole.

The bio-electric-chemicals grow
And continue, through metabolism,
Experiences, and inclinations
Forming the the life expression we become.

This is all thanks to death's prolonged sifting,
Of the rest from the best, silly from wise,
The pointless from the pointed—selection.
Oh, what ink-black rivers we had to cross!

Our birthright long signed by time, dust, and death,
Must then also serve, for Earth's living quests,
As the epitaph: RIP; time wears,
The tips of the strands rip, tear, dust is left.

SIMPLICITY

Occam sharpened his razor,
To a one-dimensional line,
Then cut his beard into strings.

They sprung from the depths,
Vibrating the songs of reality,
From which all composites sprang.

THE RAZOR

In the alphabet, Occam saw the unnecessary,
So he struck out 'j', 'q', 'x", and 'z',
Being rarities or duplicates,
And then even cut more,
Those being the vowels taking up space.

(n th lphbt, ccm sw th nncssry,
S h strck t ", ", ", nd ",
Bng rrts r dplcts,
nd thn vn ct mr,
Ths bng th vwls tkng p spc.)

But then one could only understand him almost.
(Bt thn n cld nly ndrstnd hm lmst.)

CONCISE SIMPLICITY

Writers of few words,
Even the laws that writ reality,
Say more more with less.

SIMPLETONS ANONYMOUS

Occam was running
An anonymous support group
Called *On and on anon.*

His opening remarks took but a few seconds:
"One can often say more with less."

Bill Clinton, along with many other politicians
And lawyers, then went on and on for hours.

THAT BLANKETY BLANK ZERO

Euclid and Pythagorus never even thought of it,
Perhaps not needing it for geometry;
So it was and wasn't 'Greek' to them.

Aristotle was deathly afraid of it.
Even the word 'naughty' came from it.

'0' had a chilly reception everywhere,
It's rounded symbol enclosing nothing,
As if it could be captured,
But 'nothing' never changed,
Being the same even if you took it away.

Humans stumbled on zero and nothing by accident,
Then recoiled in horror, fearing it, reviling it,
And sometimes even banning it outright,
As some kind of evil influence.

After many centuries, it seemed to be tamed,
Put in its place, as a simple little placeholder.
Then the beast reared its ugly head for real,
Misbehaving like a monster right and left:

It brought instant death by multiplication,
And wrought total absurdity through division,
Still halting our expensive computers.

It exploded into the ambiguous fog of infinity;
It ran away from us in calculus,
Sliding us down the slippery slope
Of closing in on it but never reaching it.

It spawned ghosts such as negative numbers,
Imaginaries, and those ephemeral infinitesimals.

Both the genie and the genius
Had been let out of the bottle,
And the goose egg still
Confounds and confuses,
No one knowing zilch about it,
It creating paradoxes left and right.

HALLEY, NEWTON, AND HOOKE

Halley was a sea captain, a cartographer, a professor
Of geometry, a deputy of the Royal Mint, an astronomer,
And the inventor of the deep-sea diving bell,
And wrote some on magnetism, tides,
Planet motions, and fondly on opium.

He invented the weather map and actuarial table ages,
Even proposed methods to work out the Earth's old age,
Its distance from the sun, even how to keep fresh fish,
But one thing he didn't do was to discover Halley' comet,
For he merely noted that it was yet another return of it.

He made a wager with Robert Hooke, the cell describer,
And with the great and stately Christopher Wren:
They bet upon why the planets' orbit were ellipses.

Hooke, a known credit-taker,
Claimed he'd solved the problem,
But had to conceal it
So that others could yet know the satisfaction.

Well, Halley became consumed with finding the answer,
So he called upon the Lucasian Mathematics Professor.
Isaac Newton was indeed brilliant beyond measure,
But was solitary, joyless, paranoid—no pleasure.

Once he had inserted a needle in his eye and poked around,
Far inserting the bodkin between the eye and the bone.
Another time, he'd stared at the sun for so very long
That he had to spend many days in a darkened room.

Frustrated by mathematics, Isaac invented the calculus,
And then for twenty-seven years kept it hidden from us.
Likewise, he did the same with the understanding of light
And spectroscopy, keeping it for thirty years in the dark.

For Newton,
Science was but a partial part of his life's routes,
For much of his time
Was given to alchemy and religious pursuits.

He was wholeheartedly devoted
To the religion of Arianism,

Whose main tenet was
That there could be no Holy Trinity.

Ironically, he worked as a Professor at Trinity College,
Although the only one there who was not Anglican.
He also spent an inordinate amount of time studying
The floor plan of the lost temple of Solomon the King,
Even learning Hebrew, the better to scan the texts.

Another single minded quest was
To turn base metals into precious ones,
His papers revealing this preoccupation
Over optics and planetary motions and such mentations.

Well, Halley asked Newton what the curve would be
If the planets' attraction toward the sun was
The reciprocal to the square of their distance from it.
Newton promptly answered, of course, an "ellipse".

Not finding his calculations of it
Newton not only rewrote it,
But retired for two years to produce his master work,
The *Philosophiae Naturalis Principia Mathematica*.

To Halley's horror,
Newton refused to release the crucial 3rd volume,
Without which the first two would make little sense.
There had been a dispute between Newton and Hooke
Over the priority of the inverse square law in the book.

That solved by Halley's diplomacy, the Royal Society
Had pulled out from the publication, failing financially,
For, the year before, there had been a very costly flop
Called *The History of Fishes;* so, Halley himself popped
The funds for the publication out of his own pocket.

Newton contributed nothing,
As usual, and, to make matters worse,
Halley had just taken a position as the society's clerk,
They failing to pay the promised 50 pounds to his purse,
Paying him only with very many copies of
The History of Fishes!

THE JUNE 30, 1860 SHOWDOWN

Were we descended from some ape-like creatures?
A thousand people sat down to hear the lectures.

The Bishop of Oxford, Samuel Wilberforce, rose to speak,
And, while speaking, and into his flow, looked at Huxley,
And asked if he'd become attached to apes by way
Of his grandmother's or his grandfather's recent sway.

Huxley turned to his neighbor and whispered plans,
"The Lord has delivered him into my hands",
Then rose with a relish and said something, agape,
Of the nature "I'd rather claim kinship to an ape
Than to someone using his eminence to propound
Such unscientific twaddle in a serious scientific forum!"

This was an insult to the Bishop's office and his door,
So, the proceedings instantly turned into an uproar.

Someone ran around holding up a Bible, to exclaim
"The Book, the Book!" (Truly, we'll never be the same.)

Now, who was this guy holding up the Book?
It was none other than the pilot of the Beagle.

THE CONCEPTION OF NATURAL LAWS

You cannot fool Mother Nature; it is improper,
For thou shalt not fiddle with Mother Nature;
But, since Father Time outlives all who venture,
He can fool around with Mother Nature.

At some time during a long eternity,
His paternity begot her maternity;
They then gave birth to life's certainty.

ALCHEMY HAPPENS VIA RADIOACTIVITY
AND HOW OLD CAN THE EARTH BE?

Through E=MCC we see that vast energy reserves
Are bound up in small amounts of matter, preserved.

Henri Becquerel carelessly left a packet of uranium salts
On a wrapped photographic plate in his drawer vault.

Some time later, he was surprised to discover that
The salts had burned a 'light' impression into it.
The salts were emitting rays of some sort, curiously,
So, he turned the matter over to Marie Curie, literally.

Madam Curie and her new husband Pierre, with glee,
Noted that the rocks poured out great amounts of energy,
But they never diminished in size or changed in any way.
They were converting mass into energy very efficiently.

They also found polonium and radium, and a Nobel prize,
Along with Becquerel, in 1903, Einstein yet on the rise.

Radioactive elements decayed into other elements,
Noted Ernest Rutherford and colleague Fredrick Soddy;
One day you had an atom of uranium that "bled",
And the very next day you had an atom of lead.

It always took the same amount of days.
For half of the sample to decay,
And so this steady reliable rate of decay
Could be used in kind of a clocking way.

Tick-tock, how old was it?
More than 700 millions years worth!
This age was way more
Than anyone had given the Earth.
(5 billion would be closer to the answer.)

He lectured one day,
Taking out a piece of radioactive pitchblende,
Showing it to aging Kelvin,
But Kelvin rejected it to the end.
Dimitri Mendedeyev rejected it too,
As with everything new,
Ever storming out of labs
And lecture halls all over, too;

However, the 101st element
Was called mendelevium,
In his name meant,
And quite appropriately,
For it was a very unstable element.

Pierre Curie began to experience
Radiation sickness, getting weak,
But in 1906 he was fatally run over
By a carriage on a Paris street.

Marie worked on with much distinction,
But had an affair so indiscreet
That even the French were scandalized there,
And so she was never elected
To the Academy of Sciences,
Despite not just one,
But two Nobel prizes
(Physics, Chemistry).

Scientists yet thought that radioactivity was beneficial,
Putting thorium into toothpaste and laxatives as useful;
Eventually these products were banned, by 1938,
But for Madam, who'd died of leukemia,
It was much too late.

The radiation is so pernicious and long lasting
That even now her papers from the 1890's,
And even her cookbooks, are dangerous and toxic,
So, all her lab books must be kept in lead lined boxes.
(One must wear protective clothing to look at them.)

Marie Curie was a very attractive lady, very much aglow,
For my great ancestor in his old writings such told me so.
She radiated warmth unto him as a rainbow of sparks—
"Great balls of fire!" he remarked,
"They now glow in the dark!"

FROM MATTER TO US

The big bang, or materialization,
One of many, was prosperous for us,
For its constants allowed for life's basis
'Though it didn't have us in mind at all.

It arose from some unbreakable stuff,
Perhaps several such eternal things,
Or the same from previous contraction;
But not from nothing, for how could this be?

Now, if the big bang's material result
Was not favorable for our becoming,
Then we wouldn't be here to discuss it.
'Though auspicious, it guaranteed nothing.

Matter and antimatter formed of it,
In equal parts, most of it annihilating;
However, some black holes evaporated,
Leaving a fortunate amount of matter.

Matter's here that works as building blocks,
The strong force's stability opposed
To the weak force's dispersal through decay,
Plus electromagnetism's motion.

Lucky, not planned, all this gave us a chance,
As from the stars cameth our help and hope,
When they generated all the elements
That brewed a soup of fortuitous accidents.

Earth was a golden distance from the sun—
A large number of other planets unfit;
Earth's features evolved in a good proportion
To sustain the beginnings of early life.

Soil and bacteria generated oxygen;
Death chose the useful forms over the useless,
Kept track of by RNA-like structures
In life's cradle, though we had not yet appeared.

Our blind fated path was the further paved
When asteroids finished most of the species—
Far from a feature of intelligent design,
But, it opened the space that was needed.

Evolution sifted through the accidents,
As it directed the good from the bad;
We began from the fusing of chromosomes
That made us incompatible with "chimps".

From matter to us—to all our senses,
To our brains, our minds, and consciousness,
In a universe of matter and space,
Past and future, we won the human race.

It only took 13.7 billion years,
For these many rare events to occur.
Though just a few of the coin tosses were good,
The bad tosses went nowhere in a hurry!

Well, I've left a lot of good fortune out,
And perhaps you readers can help fill it in;
Know, too, that bad luck may come as well:
Global warming, nuclear war, or more asteroids.

The lure of myth is ever great to man;
But, beyond the apparent solidity
Of the word 'faith' is its meaning—of
Belief in the unseen supernatural!

Matter and motion manufactures all
Being and time in the arena of space,
We the complex composites from simple stuff,
The ultimate, perhaps, in the universe.

OUR WORLD

Two specs of dust met and stuck together.
This was the beginning of planet Earth.

THE PURSUIT OF MERCURIA

For some years, I have pursued that lovely
Greco-Roman woman named Mercuria;
I've yearned till I could no longer reason.
Once, just her sight would have pleased me;

But now, at whatever cost and downfall,
I must taste of her fiery passion;
At whatever risk I plot her every move.
When the time is right, I'll be seeing her;

It will be just us, while the world's asleep.
The problem is that she's a fast woman
And is quite difficult to even sight,
Much less capture, entrance, embrace, and kiss.

And I can only have her for awhile;
Before dawn: if I linger with her long,
We'd soon be consumed by a rising fire;
After twilight, we'd be lost in darkness.

Yes, I have courted her many times,
But she's so elusive, fleeting, and small.
Once I waited for her just before nightfall;
All was perfect—'twas the best time of all.

There was the calm of a windless sunset,
Then the brief brooding of twilight's gloaming,
And the promise of a slow sultry night...
Clouds arrived—and so I missed her again!

She strayed not far from her fiery lover.
While I may have glimpsed her (I wasn't sure),
She slid toward her master's gravity,
Condemned to whirl about his light;

However, I was quite determined;
'Twas the thrill of the quest that kept me strong.

I planned to surprise her just before dawn...
I crept onto the frosty roof, near slipping,
There waiting. Damn! Clouds were boiling along
And blocking the view of her beauty rare.

Suddenly the clouds cleared, and she was mine—
Just over the eastern horizon was
The planet Mercury—dear Mercuria;
I stayed with her as long as possible,

Naked in the night, until, to blazes
She went when the sun arose; however,
Memories remain of those precious moments
And now she belongs to me forever.

Venus is too easy, Mars is always there,
Jupiter ever-present, Saturn bright,
The Earth under my feet, Pluto underworlded;

King Neptune, Queen Urania? Where are you?

WHAT FATE BECOMES OF LATE
THAT NEXT ARRIVES ON MY PLATE?

What has fate in store for me
And when comes the delivery?

What providence determines my destiny plucked?
Is it the stars, by chance, that rain down luck,
Or is it just serendipity and good fortune
Created by our own karma of kismet done?

What hands mold my future certainty pot,
Managing the outcome or the end of my lot?

Do the Fates 'Clotho, Lachesis, and Atropos' three
Predestine the preordination meant for me to be?
Am I doomed and bound? Is there any guarantee?
What do I care since all seems so free to be!

**FROM TOE TO BEING
AND FINDING MEANING THEREIN**

Why & How

Nonexistence can't be, nor could something
Make itself or always have been perfect,
So, before definition is the possible—
Timeless–formless—all options were open!

What, Where, Who, Then, and When

'What' matter stabilizes in 'where' space,
Begetting the appearances in motion as
'When' future moves through the 'now' to 'then' past—
This "spirit of life" granting our 'who' being.

The Forces

The strong force facilitates stability;
The weak force allows changeability.
Electric action, leading to magnetic motion,
Facilitates the movement of appearances.

The TOE to Being

The TOE has to explain origin, method,
And life, and, so, this does, the key being
That movement of appearances begets
Changes in time, showing in our life's realm.

Universal Answers

Since there's no rhyme nor reason for existence,
We're free to make our own meaning of it;
If we don't, then it's really meaning-less;
If we do, it becomes the ultimate!

Luck Happens

Asteroids swept away many species;
Two chromosomes fused, leaving chimps behind;
RNA remembers all survivors;
Good fortune smiled on Homo Sapiens.

The Balance Sheet

Life on Earth is death's borrowed debit;
We spend this life on good fortune's credit;
We're not God's puppets, but free of the strings;
Dispensing with angst, we're free in being.

We Are What We Are

Unintelligently designed, humans
Were a lucky accident of nature,
A haphazard Rube Goldberg 'invention',
With a nervous system ruled by ancient times.

The Lucky State of Us

As an accident of evolution,
We have the ultimate freedom of choice.
No "God's will"—we're beyond instinctive;
We're free to grow and evolve, through learning.

Difficulties Abound

Emotion often bypasses the intellect;
Many stand at the brink of insanity.
Only education can save the world;
We're at the turning point of history.

Wishful Thinking

Pride: Ego exaggerates self-importance
To claim that we're specially created,
Deserving a divine destiny.
Humility: we're electrochemical.

Unfortunately...

Those who can't or won't learn are doomed to stay
As their robot selves, living the sitcom life,
But, learning disperses the myths of old;
We make our own way or stagnate and die.

Meaning—or Not

Direction arrives, or one goes nowhere;
Growth happens, or one vanishes to null;

Creation comes, or reaction destroys;
Planning makes a life, or it collapses.

Coming Full Circle

Searching for the ultimate happenstance
Of how we began leads to exploration
Of within and without, a rewarding quest;
Upon return, we know the place for the first time.

WAY WAY BACK

In 1909 in British Columbia, near the town of Field,
Walcott and wife were riding horses
Along a mountain trail,
Beneath the Burgess Ridge
When his wife's horse slipped a stone,
Tipping and turning over a slab of shale.
He got down and looked;
There were fossil crustaceans unknown.

The next summer he climbed up the mountain's side,
Having traced the presumed route of the rock's slide,
And there he found a shale outcrop as long as a block,
Imprinted with Earth's ancient and tiny livestock.

'Twas from the dawn
Of life's great and complex profusion
From so very long ago—
It was the famous Cambrian explosion.

THE HOLOGRAPHIC UNIVERSE

When a tree falls in the forest
And there's no one around to hear it,
Does it make a sound?

No, for there is no ear to turn
The sound waves into sound.

Nor is there a smell, for there is no nose
For the odorous molecules to attach to,
Nor has it any color, for there is
No retina to decode the light frequencies.

What does it look like, then?

It doesn't look like anything,
For there is no brain to put it all together
By detecting form, color, texture,
Size, taste, smell, or vision.

Since the entropy of a black hole is known
To depend on the surface area of
The event horizon and NOT on its volume,
Then our third dimension MIGHT BE a projection.

A projected illusion, as in a hologram,
May still be used as it were really there
Since we can make sense of it, so to speak,
But, in truth, the third dimension may not exist.

Thus, apparently separate particles,
Like created photon pairs,
Copy the other when one is changed,
Because, in truth, they are still
The same thing in the projector room.

If the universe is holographic,
Then the tree in the forest,
Whether seen or not,
Is, at heart, an interference pattern
Brought to life only when we tune it in.

This is the mystery of the realness
Of sleeping dreams revealed:
We tune in to the interference patterns,

Whether awake or asleep,
To bring alive the reality projected.

Everything connects to everything else
Through overlapping interference patterns,
And so nothing is so separate at all, as it seems,
But is one large all-encompassing whole.

Memory, too, seems to be holographic,
Residing everywhere in the brain,
Every piece associated with others related,
Instantly broadcasting all the connections.

Every part of a hologram contains the whole,
The whole universe contained within
A grain of sand, all eternity within a moment,
The universe rumbling when an electron vibrates.

We are part and parcel of everything—
We are the cosmos; we are life; we are love;
We are all that is; we are the creator
Of the dance as well as the dancer.

Whether the past is recorded and accessible
As part of the holographic whole is not known
Or whether the other two dimensions are
Projected, as well, but perhaps we shall see.

This then is the secret of the universe,
Knowing of that which underlies all reality:
Fundamental, absolute, indestructible,
Omnipresent, indeterminate, but all pervasive.

Why absolute and fundamental?

Because it is made of one piece—itself,
And therefore indestructible, and eternal, too,
And makes up all that there is, everywhere.

THE DNA OF THE UNIVERSE

The Infinite may radiate through a matrix,
Using Information or Energy to create
The Cosmic Background antenna which broadcasts
Interference patterns of virtual reality.
(ha-ha)

WHAT NO MAN HAD THOUGHT BEFORE

Alan Guth had never done anything much before,
But soon attended Dicke's Big Bang lecture tour,
And so he'd decided to study the birth of the universe;
Thus, just like that, he developed inflation theory first.

The "Big Bang" formed 98 percent of matter spent,
But, whence the rest of all the higher elements?
What flaming forge fired carbon, iron, and more?

Fred Hoyle was a nut, much unloved, a big bore;
Working with others who often avoided him,
Hoyle came up up with imploding stars, a whim
That that allowed supernovae to generate
The heavier elements at the rate of his steady state.

This process was known as nucleosynthesis,
Causing a 100 million degree heat and mist
That sprayed new elements into clouds of stardust
That could eventually coalesce into solar systems, us.

99.9% of this mass made our sun, the rest leftover dirt,
Ever colliding, two grains being the conception of Earth,
For, in every encounter there was always a winning lump
Of these endless and random, bumping, growing clumps.

(Fowler, not Hoyle, obtained the precious Nobel prize;
Hoyle had been overlooked, but to no one's surprise.)

THE ACCIDENTAL HAPPENING

What random, unsystematic event became,
So unmethodically, quite arbitrary, so lame,
Unplanned, undirected, so casual, uncausual made
As some indiscriminate, nonspecific one bade
Of haphazard stray that erratic chance gave?

THE TRANSIT OF VENUS
ACROSS THE FACE OF THE SUN
AND
THE UNLUCKIEST MAN
ON THE FACE OF THE EARTH

Edmund Halley had suggested
That if you measured the passage of Venus
Over the sun from selected places on Earth,
You could work out the distance to the sun,
By using triangulation,
And then go on to use that calibration
To find the distances
To all the other bodies of the solar system.

These transits come in pairs eight years apart
And then there are none at all for a century dark;
There were none in Halley's lifetime, but in 1761,
Twenty years after Halley's death, the world was one.

Scientists set off for points all over the Earthly globe,
Hundreds of them, but most remained in problem mode;
Many were waylaid by war, shipwreck or sickness;
Then, too, there was much damage to the instruments.

Jean Chappe spent many months traveling to Siberia
By horse, sleigh, boat and coach, nursing his criteria
Over every bump.

At last he was near,
But swollen rivers blocked the way—
Locals blamed it on him looking at the heavens.

Guillaume Le Gentil set off from France
A year ahead of time,
But got delayed and was yet stuck at sea in brine,
Impossibly trying to take measurements
From a pitching ship.

He continued on to India,
Now having eight years to prepare
For the transit of 1769.
He erected a viewing station there,
Having everything ready on the fine day of June 4th;

But, just as Venus began its pass, a cloud slid forth
Right in front of the sun and stayed there and spent
Its time exactly for the the duration of the transit:
Three hours, fourteen minutes, and seven seconds.

Enroute to a port to head for home,
He contracted dysentery and was laid up for a year,
But then finally left the territory
On a ship that was later hit by a hurricane off
Of the African coast and nearly wrecked and lost,

But, he did make it home 12 years after setting off,
Only to find that relatives had long since sealed his fate
By declaring him dead and then plundering his estate.

The few measurements from 1761 were of no benefit;
But, luckily, in 1769, Cook had watched the transit
From a sunny hilltop in Tahiti, giving enough weight
Of information now for Joseph Lalande to calculate
The mean distance to the sun
At about 150 million kilometers.

FOREVER

*Are there as many stars
As all the snowflakes that ever fell?*

Of course there are, and more.

*Does forever ever end, dying out?
Do unbreakable basics ever wear out?*

Well, once every thousand years the Bird
Of Time flies over Mt. Everest, and downward;
On some of those occasions, a portion
Of a feather falls upon the mountain.

When the mountain has worn itself away,
The end of forever has thus arrived, that day.

FINDING THE EDGE OF THE UNIVERSE!

At Princeton University, Robert Dicke and his team
Had really been building up much scientific steam
From pursuing George Gamow's good suggestion
Of a deep space Cosmic Background Radiation.

Gamow wrote another paper suggesting some ways
To use the Bell antenna, but no one read it in those days.

Unknowing of this paper and unbeknownst to Dicke,
Arno Penzias and Robert Wilson, but 30 miles not far,
Were diligently trying to get rid of this very CBR!

At Bell Lab,
Their large communications antenna deployed
Was plagued by some persistent background noise,
A steady steamy hiss, unfocused and unrelenting,
They ever attempting
To squash it away very painstakingly.

For a year they'd tried to eliminate this nuisance noise,
Through testing, rebuilding, and wiggling-dusting ploys,
Even placing duct tape over each & every seam & rivet.

They even wiped away a ton of bird shit from the dish,
Scrub brushing it and sweeping it clean; but, no fish.

Little did they know
They'd found the edge of the visible universe:
The very first photons were at hand—
The most ancient light,
Although time and distance
Had changed it into microwaves.

It was this interfering radiation
They wished to swish away.

If the Empire State Building was the universe we know,
They had reached within an inch of the sidewalk below.

In desperation, they called Princeton about the noise;
"We've been scooped!" Dicke sadly told all of his boys.

Penzias and Wilson received the 1978 Nobel Prize,
Even though they'd not been looking, CBR-wise,

And didn't even know what it was when they found it,

Nor had they ever described it in any scientific paper,
Not even knowing the significance of it,
But from the newspaper.

(Sadly, all that Dicke's team got
Was a bit of sympathy.)

Note: they didn't really call it "bird shit",
But a "white dielectric material".

See The Birth of the Universe At Home:

You, too, can detect the ancient CBR;
Just tune your TV to a blank channel;
About 1% of the dancing static is the CBR.
When there's nothing on, it's really everything!

THE ASTEROID THAT MAY DESTROY HUMANITY

The air beneath it
Couldn't get out of the way of the rock,
Rising in temperature ten times more than the sun is hot.

Everything and everyone crinkled and crackled in the heat;
A thousand cubic kilometers of earth blew from beneath.

This shock wave, radiating at about the speed of light,
Would sweep just about everything else out of sight.

From further away, one would see a blinding light,
Then the unimaginable grandeur of an apocalyptic sight:
A rolling wall of silent darkness as black as midnight.

It would reach to the heavens,
Filling the entire field of view,
Traveling far beyond the speed of sound
Toward me and you.

A bewildering veil of turmoil
Would [ful]fill our vision
During those few last minutes
Before we met oblivion.

THE OTHER SHOE DROPS

Determinism doesn't sit well, at first;
Its flavor does not quench the thirst,
For then it seems we but do as we must,
But, we'll see a way that in this we'll trust.

We wish that our thoughts reflect us today,
Our leanings, for it could be no other way.
To know, let us turn to the random say
To see whatever could make its day.

Shifting to this other, neglected foot,
What could make the random take root?
It would have no cause beneath to explain
The events, they becoming of the insane.

We could pretend, imitating air-heads,
Posting nonsense on purpose in the threads,
But that then we meant to do this way,
Noting history, too, so random holds not its sway.

There's less problem of a determined Nature
Than the same in our individual nature,
But, sense isn't made from random direction
That relies on naught beneath its conception.

Would we wish it to be any other way?
Doing any old thing of chance that may?

The random foot then walks but here and there,
Not getting anywhere, born from nowhere.
The unrooted tree lives magically, unfathomed.
Is not then randomness but a fun phantom?

The opposite of determined is undetermined,
The scarier ghost that's never-minded.

MEASURING THE SIZE OF THE EARTH

An English mathematician, Robert Norwood, of many,
Wished to know the circumference of the Earth, as any,
With his back against the Tower of London, he forked
Two devoted years marching 208 miles north to York,

Repeatedly stretching and measuring a piece of chain
As he went forth through all the heat, cold and rain,
And made many meticulous adjustments tolled
For the rise and fall of the land and the meandering road.

Then, in York, a year since he began in London,
He measured the precise angle of the sun.
Thus, using trigonometry to size a degree of the mark,
He came up with 110.72 kilometers per degree of arc.

Not thinking that these measurements could be true,
Since the slightest errors could throw them into the blue,
Jean Picard spent two years trundling and triangulating;
Using quadrants and pendulum clocks, he got 110.46.

But, was the Earth fatter at the north and south poles?
Now new measurements were needed to replace the old.

A hydrologist, Pierre Bouguer and and soldier,
Charles Marie de La Condamine, with many bolder,
Traveled to Peru to triangulate through the Andes,
To measure the meridian from Cuenca to Yoarouqui.

They needed but to go 200 hundred miles for one degree,
But everything began to go wrong, often spectacularly.
In Quito, they provoked the locals, getting stoned away;
Their doctor was murdered and the botanist went crazy.

Fevers and falls claimed even more; the senior member,
Pierre Godin, ran off with a pretty thirteen year old girl.
Then they had to halt their work for eight long months,
Having to sort out a problem in Lima with their permits.

La Condamine and Bouguer stopped speaking,
And all the officials had many suspicions, unbelieving
That the French would travel halfway the world around
To measure the world right here in their very own towns.

Why didn't they make the measurements in France?
Well, Edmund Halley, an exceptional figure, by chance
Got from Newton that our planet was slightly oblate;
But, Jacques Cassini had come up with the reverse fate.

Jacques erred, but the Academy sent the team in mind
To South America, to mountains with good sight lines;
But, the mountains of Peru were often lost in the clouds,
So they'd wait weeks to observe a bit, complaining loud.

The terrain was near impossible, defeating the mules;
The men plodded on, fording wild rivers, hacking jungles
And crossing uncharted stony deserts far from supplies,
Tackling the task for 9 long sun-blistered years of lies.

They then found out that another French team, cold,
Had taken measurements in Scandinavia that showed
That indeed a degree was longer near to the poles,
The Earth Forty-three kilometers wider equatorially
(Than from top to bottom around the poles.)

Still not talking, Bouguer & La Condamine just moaned,
Returning to the coast, even taking separate ships home.

THE LAST DODO WORKED IN A MUSEUM

The famously flightless bird, the good old dodo,
Had a dimwitted but ever trusting natural motto.
During millions of years of isolation from us,
It had evolved on the island of Mauritius.

It was not at all ready for human behavior low,
Even waddling over to note the fall of its fellow.

In 1755, seventy years after the last dodo's word,
A museum director at the Ashmolean in Oxford,
Nothing that its dodo specimen had become "tired",
Being unpleasantly musty, so he threw it into a fire.

We are now not even sure
What a living dodo was like,
But for some oil paintings.
We will not again see their likes.

MASTODONS AND EXTINCTIONS

In the late 1700's, Cuvier could take heaps of bones
And whip them into shapely forms not in the stones.
After describing and naming
The fossil elephant the mastodon,
He put forward for the first time
A theory on extinction.

He said that from time to time
There were global catastrophes
In which some groups of creatures "became history".
This raised uncomfortable implications at the time,
For why would God create and destroy
Without reason or rhyme?

This suggested an unaccountable
Casualness by someone unseeing
And greatly troubled the belief in
The Great Chain of Being,
Which held that the world was carefully ordered for us—
And that every living thing thus had a place and purpose.

Meanwhile, William Smith noted a correlation in fossils
In rocks to find the relative rock ages that were possible.
At every change in rock strata, certain fossils vanished,
While in others they carried on into subsequent levels.

Now it was seen that God
Had wiped out creatures extinct
Not only occasionally but repeatedly—
Which made us think
Him not only careless
But having an outright hostile distinction.
There had been more than
The Biblical Noachian deluge extinction.

BACTERIA:
THE BACK DOOR TO OUR STOMACH'S CAFETERIA
AND THE INVINCIBLE RULERS OF THE EARTH

For two billion years in the Archaean world, bacteria
Were the only forms of life. Algae, or Cyanobacteria,
Learned to absorb water molecules, dining on hydrogen,
But releasing oxygen as waste; photosynthesis began.

The world began to slowly fill with "poisonous" oxygen,
But not right away, as it first combined with iron then,
Producing iron oxide that sank, that on the bottom lay,
In primitive seas, the world literally rusting away.

After 2 billion years, the atmosphere had some oxygen;
A new kind of cell arose. Some oxygen-using organisms
With organelles produced an energy much more efficient.

This was the endosymbiotic event of a mitochondrion
Which made complex life possible, by a liberation
Of energy from food, feeding on nutrients we take in.

We need them but they don't need us, for without them
We couldn't even live for two minutes.
They don't even speak the same genetic language
As the cells in which they live.

These eukaryotes are old and unknown visitors
Within our homes who've stayed on for a billion years.

In another billion years they learned to form together
Into complex multicellular beings, yet, still this world
Of the small was to ever live on and rule the world.

At dinner, Louis Pasteur used a magnifying glass for
Searching for microbes in his food, until invited no more.

There are 100 quadrillion bacteria within us & upon us,
Ever grazing on our flesh and digesting our food bus.
The Earth is not our planet, but theirs; they let us live.
They even purify our water and keep the soil productive.

A single bacterial cell can generate 280,000 more a day.
They can also share information, taking a piece away
Of genetic code from any other any time. They swim
In a single gene pool—an invincible super-organism.

They live in caustic lakes, in Antarctica, in boiling mud,
And even thrive seven miles down in the Pacific Ocean;
In sulfuric acid, too, and in a 166-year-old bottle of beer,
And can even gorge themselves on plutonium nuclear.

Bacteria were yet alive in a sealed camera lens stowed
On the moon for two years, but they seemed a bit slowed.
Some were even found two thousand feet below the Earth
Dining on what's in rocks, like iron, sulfur, and dirt.

Some frozen ones were even revived from the 3 million
Year-old permafrost of Siberia, and even one older than
The continents, was resuscitated, a 250 million-year-old
Bacterium that had been trapped in a salt deposit hold,
Two thousand feet underground in New Mexico, maybe.

SATURN'S CRYSTAL HEXAGON

On his cell phone, Graham texted a telegram,
Which is really what texting is like, damn,
With it's new codes and abbreviations
(Which is too long of a word for its definition),
On out toward the sexagon on Saturn.

Graham needed more space
From his wife-said place,
So to the ice palace hex,
He went to have good sex.

Back later came a mama-gram from the ma'ms.
Those lovely ladies there said for Graham the man
To come back, bringing quickly some quarks
For their quirks...

Came another tell-a-Graham from those ma'ms,
Saying, Master, we six slaves awaiting
Have just delivered those many sequels
Of that last and happy Saturnalia's ringing bells.

Graham went off to attend to these family matters
On Saturn concerning his sexagon,
Although he still claims that his sex is a-gone.

THE SUPER TOE IS CAUSELESS,
THUS, THAT IS THE SUPER TOE!

Our train of thought has driven us to the answer,
Of all that borne from near 'nothing' onto eternity,
Of the origin of the original disorder,
The lone dawn of our trackless radix,
Via the rails and tunnels that ever ran out:

There cannot be ever more and more
Causes beneath even more extended causes;
Therefore, intuitive or not, the causeless is,
Being such as what we observe it in the quantum.

Thus, cause is only of our higher realm,
As downward thence to its root emergence—
'Possibility' needed no mother but itself,
An egg burst open, born without a chicken.

The causeless bottom is the potential
Of possibility that is/was ever there.

Since it's 'defined' as an undefined chaos,
There's no problem of no initial definition had,
Since it can't have one and so it needs not any.

Things themselves become and go of 'virtual' potential,
Some things remaining as the rather-enduring real.
The potential is as near to simple as it gets,
Second only to the nonexistent Nothing, of course.

So, then, the potential is of no mind or 'seeing',
For that thought system can never be constituted,
As there are no more fundamentals upon more;
For, the Potential is already the ultimate basis.

Simple things ever combine, and further up,
And/or go must through phase changes,
Leading to more complex composites/forms.

Nothing, not existing at all,
And not even being able to,
But, perhaps threatening to,
Is the simplest state of all,
So, it must ever jiggle about,
Manifesting as loose 'change'.

You might say, then, that, that is exactly why
There had to be the potential for things;
Otherwise... Total Nothing, forever.

We have now reached the unexpected TOE,
One that even satisfies the ongoing trend,
For, looking down, we've always observed
The ever descending simplicity of Nature.

Now, as such, we can't really expect to find
An Ultimate Complexity sitting
Around there at the simplest point.

We didn't find Mind there;
Thus, we are ever free to be.

This causeless bottom 'fate'...
Was/is, too, a 'magical' state,
For anything could become of it.

1816: THE YEAR WITHOUT A SUMMER

In 1815, on the island of Sumbawa in Indonesia,
Tambora exploded, killing a hundred thousand people.

The Earth began to cool from the smoky ash and dust,
And sunsets became extraordinarily colorfully beauteous;

Lord Byron dreamed that the bright sun was dying;
The spring never arrived and the summer was very trying.

Crops all over failed to grow; Ireland was famished.
Earth's temperature had fallen by but 1.5 degrees F.

They called it the year of
Eighteen Hundred and Froze to Death.

FREE WILL

Do you control your thoughts or do they control you?
Could you, silly as it seems,
Just be falling, hook and line, for your thoughts?
Think about it—thoughts may tell you the answer!

The brain's decisions are determined by
Memories, associations, and
Learned behaviors right up to the instant;
So—our decisions are predetermined.

The "free" in free will has no real meaning,
Unless we take it to mean random, that
One's will depends on nothing but dice rolls;
What good would such a brain be anyway?

Can you start or stop your thoughts? In other words,
Can you will that which does the willing? Try it.
Oops, a surprise thought just came from the blue;
You did not will it—the will is unfree!

A mind is perhaps many little minds,
Each a simpleton awaiting control,
Such as when we eat, socialize, or fight,
None of them very complex at all.

The brain, with its hundred billion nerve cells,
Does all of our decision-analysis,
Only making its results known, at the last,
To the brain's highest level: consciousness.

People act, robot-like, since they know not
The why of what they do, for decisions
Are made blind, by brain networks, just before
They're presented to us in consciousness.

Consciousness comes three hundred milliseconds
After the brain does its analysis,
And, thus, has but last-second veto power,
If any, over what the brain comes up with.

Decisions are not made by consciousness,
Although, this fine picture in the mind's 'I',
Merely the brain's perception of itself,
Is fed back whole for future shortcutting.

Not much of what the brain does reaches
Consciousness, and even when it does,
The mind's last to know, being like a tourist—
For decisions precede their awareness.

First-level people have beliefs and desires,
But second-level people can have beliefs
And desires about their beliefs and desires,
Becoming able spectators of themselves.

Although our decisions of the instant are
Fully determined, and are therefore not free,
We may happen to learn something new—and make
Choices tomorrow we wouldn't make today.

Thoughts good and bad come and go, as the brain
Looks at itself without assigning values.
Still, lucky that others can't read our minds,
'Though forbidden thoughts are normal and sane.

If you try hard not to think of something,
Then you will just think of it all the more
So, if told to avoid impure thoughts, you'll
Think of people naked beneath their clothes!

We may fall for our thoughts, hook, line, and sinker:
Conditioned responses, reflexes, or
Overwhelming emotions, spurious,
Or ancient, planted by evolution.

When extreme thoughts arrive, uninvited, as
Most thoughts do, we veto them, saying "don't",
For while we can never will that which does
The unconscious willing, we have some "free won't".

We're all robots, but no one notices
Since there are so many different kinds,
Which, though making life quite interesting,
Obscures the fact that the will is unfree.

AN UNREAL EXPERIENCE

I climbed the Himalayas, long ago,
And complained to the wise Lama up there
That life could be hell.

He said "Get lost!
Go make a heaven of hell and then me tell.
The door is never shut on the prison cell,
So, why would you ever want to stay inside it
When the door is always so wide open."

A week passed, then a month, and then 30 years,
And I found myself at a Buddhist-run cafe,
And decided to sit there
Through most of the summer,
Having just retired from IBM
And becoming as free as a neutrino.

The cafe was in New Hamburg, NY,
And was run by the Buddha Girls
From the monastery on Shafe Road,
Home to one of only two Lamas
In the entire United States,
And the only one on the east coast.

The cafe was called
"Himalayas on the Hudson",
And the Lama often came to dine there,
With his entourage of higher-ups and bodyguards.

Because I was there often,
I got to know the old Lama,
His bodyguards ever retreating,
And so I taught him how to do
High fives and low fives and such,
And we began to talk about
The connectedness that underlies all things,
The reaching of which meditative state
Through the removal of all thoughts
Being the very heart of Buddhism.

In addition, I always gave him the weather
For the rest of the day and for the next day,
Always saying that
It would become sunny if it was raining,

And that it would be still sunny
If it was already sunny,
And if it was really raining heavily,
That it was always sunny on the inside.

I remember,
Thinking upon first meeting him
That "here he is", the great one,
And so I have a chance to ask
A deep question of him,
Without having to go back over to
Tibet or India and climb up a mountain,
So, I pointed to an article
In the newspaper that said,
"We may never know who won
The Presidential election, Bush or Gore"
And so I asked him for his wisdom on the matter.

Well, he thought for only a second or two and said,
"Who cares!", and such it sunk into me a bit later
That this was a great wisdom, indeed.

The Cafe workers didn't wear the flowing gold
And reddish robes that the visiting Buddhists wore,
But wore regular clothes and even had long hair,
And so, many of the hectic type customers,
Unknowing of their servers' Buddhism,
Wondered at the peace and joy that the workers radiated,
As if they were in some sort of serenity field,
Which I suppose the workers were,
Plus being chosen for their outgoingness.

I talked with them about string theory,
The theory that the differing vibrations
Of really small 'strings' gives rise
To all of the elementary particles and forces,
And, so, we related this to all that is absolute
And fundamental beneath this projection
Of reality in which we live out life dreams.

Buddhism is not a religion, but a way of life,
And they can still have friends,
Outside jobs, fun, and whatnot,
Although some of them spend a lot of time
On the inner world, which, like meditation,
Can only be known as "not what you think".

Summer soon died in his sleep one night,
And so Time hurled its waves ever onward
Until even Old Autumn had passed on.

The cafe had been rented out
And had now become
An American-Korean restaurant
Run by Sin-Ha and Su-Nee,
Although still owned by the Buddhists.

Winter had snowed us in.

In late spring, the Cafe, still my 'office',
Announced that it was closing down,
Right away, for it could talk,
Although its Garden of Peace and Serenity,
Surrounded on three sides by 30-foot rocks,
The "Himalayas", was still open,
And so I figured that it was time
To move my "office" outdoors,
Not that I would ever do any W-O-R-K there,
For that is a four-letter word to a retired person.

Then, miracles of miracles, that day,
After saying good-bye to the Koreans
And taking home 50 eggs
And many bags of chocolate chip cookies,
I went back to the Cafe garden
To sit under an umbrella table in the rain,
And there was the old Lama himself,
Sitting there all alone,
Having just shown the building
To someone who might lease it.

I hadn't seen him in six months,
For he had been off to other continents.

He gave me a medium high five
And I told him that the sun would be out tomorrow,
And that it was always sunny on the inside.
He said, *"Thanks, old friend."*

"Re-leasing the building?"

"Yes, probably, but we'd like sell it.

Perhaps Buddhists shouldn't be in business."

"Well, it worked as a kind of outreach,
When you ran it,
And the Koreans liked it for a while."

"True."

"How's the new golden temple going?"

*"It's about half completed.
We need another three million dollars."*

"Hmmm."

*"Yes, I know.
Perhaps Buddhists shouldn't be looking for money,
Nor building a golden temple that's not really real."*

"Yes, I've heard that this world isn't really real,
That we shouldn't worry about the rain
Or about life's tribulations."

*"That's what we believe.
Tell me, does that work?"*

"Well, um, does not life's existence
Look, seem, and act just the way it would,
In every detail, as if it were really real?"

*"Yes, indeed. Exactly.
That's what they say makes for the illusion."*

"I hate to say this,
But a 'difference'
That makes no difference
Is no difference."

"I think you're onto something."

MEDITATION

People can't usually ever see
Further than an order of magnitude
Beyond where they are rutted, but...
Some can intuit ultimate reality!

It says, in those 'dreams', *Of ever waking,*
It's hard to convince you with dream-language,
As when, in wakeful reality,
To tell you of that which is beyond telling.

During meditation, one clears the mind,
And so, then, there's no real self, just one quale—
A near nothing that has little need to be;
Is this what-it's-like to be a pure soul?

Physics, once more direct, seems now but an
Immaterial science of math-shadows,
While mysticism, once but a foggy notion,
Now's the direct observation of the Light.

Meditation shifts intention away
From controlling and acquiring,
Toward acceptance and observation:
You take-in instead of acting upon.

Enlightenment's not grasped or possessed—
Acquisitive aim locks the secret out—
The form of consciousness that one starts with;
This is why "the secret protects itself".

The 'spiritual' refers to profound connection,
Though not through visions or ecstatic emotion,
But with the experience of connectedness that
Underlies reality, and nothing more.

Meditation relieves the survival self,
Shifting attention from acting to allowing,
From emotional identification to observation,
From instrumental thinking to receptive experience.

Meditation, renunciation, and service
Are not really mysterious, just different
From the usual object-oriented approach.
Mysticism is modern and ancient, not esoteric.

In serving the task, one forgets the self,
And accesses life's connected aspects
That go beyond one's self-centered consciousness—
The survival of mankind being at sake.

Awareness is the ultimate being,
Fundamentally connected with 'soul',
And cannot be known in terms of worldly
Objects—it's like, well... you have to be there!

The connectedness of everything to everything,
A rudimentary perception in and of itself,
Experiential in its ultimate physical disposition,
Facilitates our consciousness of exterior through interior.

Not exactly. Actually, the quietus
Of the brain's self-boundary & ID center,
Via focus on mantras, hymns, or prayers,
Is but a neurological effect, nothing more.

(Tested via electrodes in Buddhist monk meditators)

THE CURIOUS ENCOUNTER WITH MADAME

De Broglie declared that all motion
Of particles must be associated
With the propagation of a wave.
Einstein then wrote that De Broglie
"Had lifted the corner of the great veil."

Einstein later had an opportunity
To lift another veil—that of Marie Curie,
When they vacationed together,
Quite reactively, in the Swiss alps.

Did they or didn't they exchange energy?

Einstein was a ladies' man, though married,
Busy having an affair with his cousin, Elsa,
And Marie was a married man's lady (Paul's).

Einstein wrote his wife that Marie was a grouch,
But was this just a misdirection meant to allay?
They inhaled the alpine air, talking science,
Strolling far and trying to name the peaks.

A STAR IN THE DARK

Old Junius, the astronomer prime,
Had been saving up his star gazing time
On the Hubble, on the condition that
They would let him use it in one long shot,
That he needn't reveal his gamble's plot,
And that they could use whatever he got.

Time was precious on the great starry stage,
But Junius had stored up seven days
During the last ten years, yet needed more,
Although he was now ninety years and four.

The Director gave in, made it eleven,
Wished him well, and gave all a vacation,
Instructing them to return here in ten.

...

They all came back, with a long day to spend,
Sitting around, wondering what the end.

One noted, "I heard that the scope began
In the dimmest area of the sky,
And it didn't even move after that!
There're so many better places to look."

Another said, "Hope he's not gone senile."

The dawn broke, and old Junius came out,
Clutching a bottle of wine and a cigar,
Although he neither drank nor smoked.

Junius began, "I looked at empty space..."

They began to squirm and shift in their chairs.

"...At a coal black spec no larger than a
Grain of sand held out at an arm's length,
Which is one thirteen-millionth of the sky."

They were at a loss for something to say.

Junius began to open the wine bottle.

They all looked to the Director, who smiled,
Noting, "You let the light accumulate.
Any questions for Junius to articulate?"

One ventured, "You found a star in the dark?"

"A star?" Junius said slowly, "A star?"

He took a sip of wine and lit the cigar.

Someone got alarmed, "You can't smoke in here!"

"I can today," Junius answered,
"I found 10,000 galaxies afar,
Each one of them having a trillion stars."

They all applauded then sat back speechless.

The Director stood and pulled out a note,
"I've borrowed a few words from Blake the poet:

'To see a World in a Grain of Sand
And a Heaven in a Wild Flower,
Hold Infinity in the palm of your hand
And Eternity in an hour.'"

UNIVERSAL NATURAL LAW, AND THEN ITS VIOLATIONS

You will always be caught,
So don't even give it a thought.

The violation of universal natural law
Is the cause of our problems, all,
Of everything that becomes rife
That plagues individual and national life,
These stresses only leading to more strife,
From lowlifes leaving their wife, for the wildlife
Of nightlife, to cutting someone with a knife.

So stem problems of national health,
Crime, the economy, education, wealth,
And the black environmental sins,
All of them having their origin
In a widespread law violation
By some portion of the population.

Universal Natural Law is very terse
In governing the entire universe,
It being the orderly principles
That regulate physical events/processes.

Science defines the universal law of nature,
A precise description of how nature matures.

Universal law pervades everything,
Of all that is in passage and being,
From the motion of particles
To the evolution of life's articles—
Operating at every scale:
The subatomic, atomic,
Molecular, biological, geological,
Astrophysical, and cosmological.

The universe is structured, hence,
In these many layers of existence
As worlds within worlds,
Distinguished and not only furled
By vastly different time and distance scales,
But that every level has its own set of details;
For example, an electron/nucleus system
Is not analogous to that of a planet/sun.

The more superficial macroscopic levels of nature
Can be seen as fragmented expressions, for sure,
That are manifested from the more unified laws
Governing deeper levels with their scrimshaws—
The reflections of the dazzling symmetries
Of what once were inaccessible mysteries.

The outer 'becomes' are based on inner ones,
The only fountainhead of all the rhythms.
And the converse is not true.

Nature's governance is maximally efficient,
For it is frugal, as not a spendthrift—
It following The Principle of Least Action
In all of its action and protraction.

This is why a ray of light refracts
When going from air to water's tract,
Minimizing the time
And saving every dime.

From this maximal economy of nature,
All classical behavior can be scriptured.

Entropy is a count of quantum states
Accessible to a macroscopic system's estate,
This available number ever increasing;
The nature of life is to grow, ever reaching.

The path of least action's welcome
Is just the macroscopic outcome
Of the simultaneous superposition
Of multiple coexisting paths' auctions
At the microscopic level,
The outcome ever of the least income.
The law to which all must succumb.

All is rooted in the verses
Of the Constitution of the Universe.

Life takes advantage and cause
Of the universal natural laws,
Even such as in merely walking,
Which is an immensely complex undertaking.

We employ technology
In all of its variety.

Everything that we fail to accomplish
Is but due to the total failure
To apply universal natural law effectively,
This being the source of all difficulty.

In the absence of knowledge of a lever,
The simple task of moving a boulder
Becomes complex and arduous to the shoulder.

Not learning gravity has caused non-mild
Injuries to many a young child;
The old uses of radiation caused cancer tumults;
The use of DDT had many adverse results.

Smoking cigarettes, heavy drinking, being out late,
And other addictive obsessions surely violate
Universal natural law, at whatever rate,
Resulting in negative consequences,
While psychological violations dispense
Stress directly in a sequence immense.

While fulfillment of desire can bring happiness,
It also raises the scope and standardness
Of future desires, making the duress
Of frustration an inevitable process.

Over time this causes psychological stress,
Which in turn impairs creativity's success,
Stalling future desires
By watering their fires
And also leads to problems of health,
These then causing further stealth
And violations of universal natural law—
Resulting in the nonsense
Of a life out of balance—
Leading to aggression, anxiety,
Impulsive violet behavior, hostility
And substance abuse—
A vicious cycle of refuse
That, among other effects,
Fills up the prisons to correct.

HIGHER CONSCIOUSNESS

The three lower consciousnesses that are
Obsessed with the securing of objects,
With the chasing of sensations, and with
Power/control will never ever be enough.

There are NO actions of people that can
Justify our becoming irritable
Angry, fearful, jealous or anxious if
We give them our unconditional love.

If we don't accept the unacceptable,
Then we lower our level of consciousness
Our response will mirror their uptightness—
Which can spread the bad moods onto others.

Conscious Awareness, which can but witness,
Is a safe haven from which to observe
The drama of our lives playing in our minds,
Granting us a sobering distance from it.

From a safe subjective place that's free of fear,
Our soul, our conscious awareness, can witness
The strange thoughts and emotions that surface
On the mind, sent by the subconscious brain.

Putting ourselves in the place of others
When hurtful things are done to us,
Expands our consciousness, compassion, and love
Since we can come to know why they did it.

When we converse with ourselves, it is our
Higher Consciousness—our Conscious Awareness
Or I, that questions our lower consciousness
Impulses toward securing, sensation, and power.

Seeing the big picture of life and its stages
And connections lets one not get annoyed, say,
At being cut off in traffic, for s/he
May be old, learning, lost, growing, or angry.

Putting the needs of others ahead of
Our own produces the byproduct of
Happiness and reduces stress, for we
No longer have unrealistic expectations.

FUNDAMENTAL POSSIBILITY

Time, space, stuff, change, and form were real-ized from
The Fundamental Possibility,
Becoming our penultimate reality—
One possible from all probabilities.

Our reality came not from nothing,
But existed always as possibility,
One that amounted to something workable,
Among all in superposition.

No form of our penultimate realness
Could have existed alone before
Everything's options were known-all-at-once,
For what could have made the choice among many?

Nor came it from an absolute nothing,
Since there can be no such "thing" at all,
So, since either way is impossible,
Fundamental Possibility <u>is</u>.

This ultimate basis of reality
Though not much like our local reality,
Is hinted at by quantum physics—
It forms reality real as can be!

So how else could it be, for particles
Do appear and disappear from somewhere,
Going from here to there with no between,
Manifesting from no-where to now-here.

I'll follow every single avenue,
Whether it's brightly lit or a dark alley,
Exploring one-ways, no-ways, and dead-ends
Until I find where the truth is hiding.

Since we all became of this universe,
Should we not ask who we are, whence we came?
Insight clefts night's skirt with its radiance—
The Theory of Everything shines through!

Some simple substances gave rise to everything,
Chosen as probable above the rest—
Known all-at-once that it would be the best—
The most promising—the possible ones.

As to how complex, there is no limit
But to collapse into a black hole;
The smallest of all is the planck distance,
So size is absolute, not relative.

Like the moon, challenge night and gain the light;
Like the rose, suffer the thorn—gain the fragrance;
Of life, surrender to live forever—
Enlightened more than a thousand suns.

World does not pass by—you pass through it;
Clear your being so the treasure may arrive;
This spirit sparkles of a different light—
The gemstones are of a different mine.

THE AGE OF RUST

Aliens visit the Earth, thinking lush,
Only to find it covered with rust.

Their visit does not occur in future times
When the Earth is full of dusty rime,
But was way back in the youthful years
Before there was any atmosphere.

Bacteria have just begun making oxygen,
Discarding it as a mere waste product—
An unwanted poison to be jettisoned.

Everything on Earth that is capable
Of being oxidized becomes oxidized;
We've seen evidence of this rust
In bands of red oxide deep in the crust.

Only when this rusting comes to an end
Do the oxygen levels in the atmosphere rise;
They are only 1 percent after 2.4 billions years;
Now being 20 percent after double those years.

The aliens soon leave our planet,
That rusting junk pile that
Could never amount to squat.

They did take a photo of the Eiffel Tower, however.

THE FISH WHO ALMOST EVOLVED MORE

On the road to Kingston, NY, the other day
I was happy to see that the rains had returned
And that the drought was ending.

I stopped to do a little fishing along the way,
And it reminded me of a fish
I'd caught during the dry season.

I'll tell you about it now.
It took me awhile to get over it.

It was so dry that I could walk
Across the reservoir and the creeks.
The water was shallow due to the drought,
And most of the fish were swimming sideways
So they could stay under water.

Then I saw a rather amazing sight:
One fish was leaping from puddle to puddle,
Sometimes crawling across the dry land in between.

I threw away my fishing pole
And caught this fish with my bare hands,
Thinking of the delicious fish fry dinner
That I would have that night.

I put the fish in a bucket of water
In the front seat of my car
To keep it fresh during the long drive home.

Every so often that fish would poke its head
Out of the water bucket and look at me,
Sometimes even trying to jump out.

Finally, it did get out of the bucket
And sat on the seat next to me.
It was then that I realized that
I could never eat this fish.

About the same time, a brilliant idea stuck me:
I would train this fish to live out of water,
And make a pet out of it,
As the fish seemed to already
Have inclinations in that direction.

At home I put the fish in a barrel of water,
And sure enough, it tried to jump out.
So, each morning I would take it out
And put it on the grass,
Which was still wet from the dew.

Then, when I could see that it had had enough,
I would put it back in the barrel.
Each day the fish seemed to last longer and longer
Outside the water barrel before getting listless.

After a few months of this,
The fish didn't need much water at all;
As I walked along the road in the morning,
It would wriggle along beside me
In the wet grass in the shade.

Later, when the day became really hot,
I would give the fish a drink from my water jug.

After a few more months of training,
The fish was able to flop and sort of 'swim' along
Down the middle of dusty roads.
And when I offered it a drink, it refused!

We even went to the beach together;
Of course; only I went swimming—
The fish just laid on the sand, getting a tan
And enjoying the breeze.

One day it was over 120 degrees
And the fish just had to have a drink,
So I gave it a dry beer.

Other than that,
The fish never touched water anymore,
Having become a land animal.

What a lovable pet!
It slept with me, saw movies with me,
Went out to parties with me,
Chased down tennis balls
And brought them back to me,
Rode on the back of my bike, etc.,
We were inseparable!

But then a tragedy happened:
We were walking down the road together one day,
And passing over an old bridge;
Suddenly my fish fell between some loose boards
And on down into the creek below and drowned.

THE CALDRON THAT ALMOST
BREWED HUMANITY AWAY

At Toba, in northern Sumatra, a supervolcano
Erupted only seventy-four thousands years ago.
Six years of volcanic winter followed this eruption,
Bringing pre-humans to the very edge of extinction.

There were but a few thousand of them left around,
Since very little light could reach the dusty ground.
It took twenty thousand years for them to recompose;
From this handful of hardy souls we humans arose.

In 1960, Bob Christiansen looked around everywhere
At Yellowstone National Park for its volcanic caldera,
But found it nowhere. By some coincidence, NASA
Had photos from a recently tested high altitude camera.

Astounded, Bob learned
Why he'd failed to spot the caldera;
It was virtually the entire park,
2.2 million acres of area!

Yellowstone must have blown up with a violent misery
Far beyond anything known throughout our history.

The crater was forty miles across. The cataclysm was
Even beyond the scale of what the imagination does;
It had thousands of times more monstrous molten fire
Than Mount St. Helens. Krakatau was but a firecracker.

Yellowstone's eruptions average
One really massive blow every 600,000 years,
The last one being 630,000 years ago;
It is long overdue;
Better take out some no-fault insurance.

THE NIGHTMARE

She, looking like Melanie of ToeQuest,
And still in her pajamas,
Grabbed her purified water bottle
And hopped on the bus.

Glad to see her,
I waved her over to my seat,
For she was my guru.

I was also in my pajamas,
For this made the yoga
Of our meditation therapy easier.

"This is not a real bus,
Nor is is really moving," I offered.

*"True," she replied, "we are dream characters
Of the Perfect Awareness."*

"It just all plays out in our consciousness,
Kind of like a movie."

*"Yes, nothing comes through the senses
Or from the brain or any thing like that;
It's life's soap opera channel
And there is no remote control."*

"It goes as it has to go; all is illusion,
But we aren't fooled at all."

"No, we are foolproof."

"Why are you wiggling all around?"

*"I have to pee; it's a dream pee,
But it is still a dream hurt."*

"Well, you could get off at the next stop
And go into a building."

"OK."

"I'll tell you a short joke
Before the stop comes:

If you are Russian before
You get to the bathroom
[water closet to you]
And you are English
After you come out,
Then what are you
When you are in the bathroom?"

"I give up; the dream didn't tell me the answer."

"European!"

"Ha, ha. Now I really can't wait for the next stop!"

"Just ask the driver to stop near a building
Or a gas station and let you off."

"OK, I'll see you in a while."

She had to pee so bad that
She ran straight for a building
And rushed in right to the bathroom
Without anyone even noticing;
Nor did she notice what
The name of the place was.

When she came out of the bathroom
She was English again.
An orderly stopped her, restraining her.

"I'm sorry. I had to go."

"You need permission for that.
Now let's get you back to your room."

"What? I don't have a room. Where am I?"
"Why, of course, you are a resident
Of Chesterfield Mental Institution."

"No I'm not. I just got off a bus."

"We hear those kinds of stories all the time.
Where's you room?"

"I am sane," she said with a dry mouth.

"Would you like a drink?"

"No, I only drink a special kind of water."

"Oh, a special kind?
Then maybe it is in your room."

"I left it on the bus."

"There's no bus stop here.
Let's get you out of the lobby."

"I don't belong in this place."

"Then why are you wearing pajamas?"

"For meditation therapy."

"Therapy? Well, I can get you to that."

*"You don't understand. I am normal.
My God, what a turn this dream is taking!"*

"A dream?"

"Yes, nothing is real; all is a dream."

THE UNIVERSAL ACID

As a boy in Dan Dennet's chemistry class,
I wondered, as did many,
About the following scenario often dreamt of:

I mixed two compounds, which, unfortunately,
Produced the ultimate acid.

Nothing could contain it.
It quickly ate though the container,
The floor of the laboratory,
And then even all the way through the earth,
Eventually sloshing some poor sap in China.

This, too is what happens to us, through education,
As our chemical-bio-electric nature is revealed to us,
Like some kind of giant shock,
After which we will never be the same again,
As perhaps some are now reeling from,
Well, maybe just a little bit.

The biochemical mush that is us,
When fully realized,
Leaves us stunned and astounded.
We grasp for what we once thought we were before,
But, it eludes us in the new light of learning.

The universal acid of such knowledge
Eats through all superstitions, folk tales, and myths.
Nothing can contain it.

We may come to even regret
Our learnings of this condition,
For it dissolves our container,
Leaving us floundering in the lurch.
It happened to me, too, beginning in fifth grade.

But, wait, it's not so bad, is it,
For what we are is what we are,
And we still have feelings, personality,
And more adventures of learning that await.

The light of education ever shines brightly.
Many dark alleys remain to be explored,
Given our new insight into the human condition.

TO THE ENDS OF THE UNIVERSE

I took a road trip
Through the universe recently,
Smoking some pot
And playing the radio loud.

Holy-moly, there's nothing holy out there.
In fact, it's a very uncongenial place for life.
I'd much rather be in Australia.

96% of it was useless dark energy and dark matter;
The rest was mostly rocks gases and dust;
Dangerous radiation zapped all over the place;
And it was fricken freezing!

Oh, what I would have given to be in Canada.

Whatever designed the universe
Certainly didn't have life in mind;
It even took evolution billions of years
To fine-tune us to the Earth.

Then we nearly got wiped out
By huge disasters right and left,
Even once shrinking back down
To a population of around 2000.

I saw the graveyards of the stars
And some stellar nurseries, too;
All kinds of energy swirled about—
When it wasn't exploding and wreaking havoc.

I stopped to eat at the restaurant
At the end of the universe,
On a moon,
But, it had no atmosphere,
Plus, all the food had been microwaved,
By the CMBR.

What a wasteland of a wilderness of wilds
Of a whole bunch of crap
That nearly went on forever in every direction;
This was as much of a place
Unsuited for life that there ever could be.

I'm back, thank my lucky stars,
Noting that, 14 billion years
After the initial chaos, here we are,
Having beaten the odds.
Well, someone had to!
We won the universal lottery jackpot.

Oh cripes, here comes a humongous asteroid!
Darn, all that luck for nothing.
Double '00' has come up.

It was only a matter of time.

EINSTEIN AS A NEAR TRAFFIC FATALITY

George Gamow told in his book,
'My World Line',
How he was conversing
With Albert Einstein
While walking through Princeton
In the 1940s.

Gamow casually mentioned
That one of his colleagues
[Pascual Jordan]
Had pointed out to him
That according to Einstein's equations
A star could be created out of nothing at all,

Because [at point zero]
Its negative gravitational energy
[mass defect]
Precisely cancels out
[is equal to]
Its positive mass energy
[rest mass].

"Einstein stopped in his tracks,"
Says Gamow,
"And, since we were crossing a street,
Several cars had to stop
To avoid running us down".

HENRY CAVENDISH

He was a century ahead of his time,
On electrical conductivity, for example,
But kept everything so secret
That much of it didn't come out
Until the century had passed.

Without telling any one,
Cavendish discovered or anticipated
The law of conservation of energy,
Ohm's law, Dalton's Law of Partial Pressures,
Ricther's Law of Reciprocal Proportions,
Charles's Law of Gases,
And the principles of electrical conductivity.

And that's just some of it;
There was also tidal friction,
Atmospheric cooling, freezing mixtures,
Heterogeneous equilibria,
Clues to the noble gases,
And of course the gravitational constant,
And the weight of the Earth—
From a nautilus looking machine of weight,
Counterweights, pendulums,
Shafts, and torsion wires.

He was so shy that it was known
That on no account was he
To be approached or even looked at.

Those who sought his views
Were advised to wander into his vicinity,
As if by accident,
And, to "talk as it were into vacancy".

If their remarks were not worthy,
An actual vacancy would quickly appear.

NEW LAWS

These are the new laws and formulas
That reflect my attempt
To convert to Christianity...

Force of Gravity

=

SomeFactor*[(Mass1*Mass2)/Radius squared]

(law of gravitation)

(+ motion added)

This law, discovered by Newton under a tree
That is still bearing falling fruit,
Along with some other law of his or someone's
About a body in motion staying in motion
And a body at rest staying at rest,
Each of their own accord,
Was thought to cause planets
To circle a star (sun),
Although they did so as an ellipse,
For which there is also
Some kind of wonderful formula,
And so it is that if a planet had no sun
Then it would just go in a line, not ellipsing,
And, if it had a sun and it were at rest
It would plunge straight into the heart of its sun;

But, since being in motion and having a sun
The planet then takes the in-between path,
As it literally falls around the sun,
But, who cares,
For none of this is true anymore.

In actuality,
God guides the planets safely around the sun
Through their orbits, so...

GM (Gravity&Motion) = GGH (God's Guiding Hand)

Many textbooks will have to be changed
To show this new truth.

Perhaps we could refer to it as

Isaac: Revelations II.

H_2O
(Chemical formula for water)

This is the most often used formula.

It is thought that having
Such a small molecule of hydrogen
Attached to such a large molecule as oxygen
Causes the sliding around
That makes water molecules so very slippery.

However, it is really that God micromanages
Every single atom
And also even what is inside of them.

Besides, God's son showed
That water could be transmuted, so...

$H_2O => WINE$

Jesus thought of opening a brewery,
But then had another calling.

$E = MCC$
(Einstein's conversion of mass to energy)

A little mass makes for a gigantic energy
Such as that of a nuclear bomb,
But, really, although bombs work,
Pay no attention to Einstein,
For, actually, it is
That God's very powerful Energy
Makes our energy.

There is no need to worry about
Where God's tremendous energy comes from,
For God is actually a semi-unified twin-genii
Split into Good and Evil,
Positive and negative,
And so He is zero on balance

At the end of the day.
The same for good and bad angels.

Zero=Good+Evil to the infinite power.

A squared + B squared = C squared
(Pythagorean theorem for a right triangle)

No one cares about this any more,
Nor even any Geometry,
For God can make a square circle.

From now on this formula
Refers only to the Holy Trinity.

God = 3 = DoAnything

Very Complicated Formula
(radioactive decay of uranium)

This has been replaced by God's power of alchemy,
For that's what it really was anyway.

Some kind of -b+-
some square root thing over 2a, etc.
(solution of quadratic)

By memorizing this great formula
That I have now totally forgotten,
I was able to do very well in advanced algebra.

This formula will be of no use
To us in advanced theology,
Which, as in all of my seminary courses,
Will just say that 'God did it'.

The Shortest Course on 'Whodunit' Theory

I took a really short course called GOD 101:
It only lasted about 3 minutes,
And there was no continuation of 102, 103, etc.

The instruction consisted of but one statement:
'God did it.'

Note that this statement
Not only puts the answer before the inquiry
But that it also halts all inquiry;
Thus, the case is closed before it can open.

Really, God is only a theory,
But I didn't let on,
Answering 'God did it'
On the 1-question test of
'What is the answer to anything?',
Thereby passing the test,
The course and the college.

The metronome graduates
Never said 'God is a theory',
But ever produced a regulated aural pulse
Of a steady tempo, in saying:
'God did it.'

The mind makes funny shortcuts
For what it cannot know;
But, I am not one of those minds.

A—T, C—G
(four building blocks of DNA with their matchings)

It has been found
That these letters spell 'God' in old Hebrew,
So, that is the story of that one.

A bunch of functions that I hate
(fundamental theorem of calculus)

I never liked figuring out all
Of these unholy moving things,
But now I am unmoved by all things holy.

V=RI
(Ohm's law)

Ohm has been found to be ooommmm,
The focus of meditation—
On God or of a sleeping dreaming Brahman.

Pm = Po(some crap)
(compound interest)

An accountant will redo this one,
But it has to do with
Accumulating treasure in Heaven.

TOE

This equals the Ground-of-Determination.
Note that the acronym for this is GOD.

SOUL
(invisible appendage replacing the brain)

It stands for:
Spirit-Of-Unconditional-Love

Quark

A marking was found on this material:

Holy stuff:
Made by God

Evolution

Explained by anagrams:

Outlive On
Olive Unto
Ovule Into
Vile No Out
Vile On Out
Live No Out
Ovule It On

Love I Unto
Love In Out

=

No words can describe this symbol,
For it is what it is.

"WHAT IT IS LIKE TO BE ANOTHER CREATURE"

What would it be like to be another kind of creature?
Do I-thoughts of self-consciousness emerge
From their integration of experiences?

Who knows about the 'I', but, in a sense,
We already know about being other creatures,
For we already have been during our development:

The growth of a human parallels and recapitulates,
At a vastly accelerated rate, the evolution of life.
We start out as a single-celled organism,
Much like an amoeba or a bacterium.

Then we progress through the phase of a blastula,
A simple, undifferentiated multicellular stage,
To become an embryo barely distinguishable
From those of many other animals, even
Including those of reptiles and amphibians.

For these first few weeks after conception,
We are truly a lower form of life ourselves,
Bathing, as long ago, in the warm amniotic sea.

So, how did it feel to you? Can you recall?
No, for you were not around at the time;
There was no developed conscious sense of self.

But, now we do feel like someone,
Having inner depth,
Our activity and actuality
Being one and the same.

GOD ON TRIAL

"Jehovah's" trial for crimes
against humanity begins thusly,
but ends well:

"Do you, God, swearest to tellest us
the whole truth and nothing but the truth,
so helpest you God?"

*"Which scriptures of what bible should I swear on?
There are so many."*

"Oh; here's a Mormon bible,
with a whole extra section
that was transcribed from
the golden plates You sent."

"I didn't send those plates."

"OK, let's not worry about that now;
We'll come back to it later.
You are truthful, are you not?"

"I can do no evil, and that includes not lying."

"Finally, a believable defendant.
What is your full name?"

"'God Damnit' is what I am usually called."

"Ha-ha, but what is your real and proper name?"

"None. I am what I am."

"Um, any aliases, like Lord, Jehovah,
Almighty, or such?"

"No."

"Are you sure?"

*"Yes, those are just some names
that people call me, plus even very bad names."*

"But you do exist as you are?"

"Depends on what the meaning of 'exists' is."

"You know, like 'to be', being One that is."

"Depends on what the meaning of 'is' is."

"Is that your lawyer, Bill Clinton, sitting over there?"

"Yes, for he can get out of anything."

"But is he going to talk endlessly in your defense?"

"No, for he has been going to 'On and on anon'."

"Good, now how come
we can hear you but we can't see You?"

"I am invisible, plus, you are like schizophrenics."

"Hey, no name calling, order in the court!"

"I'll have a cheeseburger, no pickles, no onions."

"That's more like it.
So you mean we are just hearing voices?"

*"Yes—do you remember the study
that showed that 17% percent of priests are schizophrenic,
but only 1-2% of the general population is?"*

"Oh, yeah; but You're not getting off that easily."

"I am innocent."

"What did You do before You Created things?"

"I was being made Myself by Myself."

"How did You do that?"

"Recursively."

"OK, anyway, did you have intercourse
with a teen-age virgin?"

"Heck, no, she was underage;
I only date 30 billion years old women."

"Still single?"

"Yes, for as Mr. Always Right
I just couldn't find Miss Perfect."

"So, Jesus was not Your son then?"

"No, but he was a really good guy—
a human telling stories that everyone liked to hear."

"But, anyway, you are a 'He'?"

"So they usually say."

"Don't You know?"

"No, for humans created Me in their own image
and with their own traits, so I am male."

"Are You jealous of any of their other imaginary gods?"

"I am above all that lowly human-type emotion stuff.
I am Perfectly Good and absolutely totally full of Love."

"Love is a human emotion."

"Yet that is the only emotion I have,
for it is the ultimate one."

"So, You never do evil?"

"Depends on what 'evil' is."

"Well, as in things like harming others,
except in self defense,
stifling the growth of mind,
and creating false ways of living,
arbitrarily, through use of imagination
of what the concept of good 'should be', absolutely."

"I am not capable of evil. I detest evil.
I would hate Myself if I performed any evil. It is unthinkable.
Then I would be in the category of a devil."

"Is there a Devil?"

"No, I would not tolerate any such thing,
for then it would sway humans to sin."

"You appear to be without fault,
but we still have to continue this trial."

"Thank you, but I have no-fault insurance."

"Did You do away with almost
everyone on Earth with a Great Flood?"

"Heck no, human nature is exactly the way
it is supposed it to be, as is.
What do you think!
God not a big fat goof, that is,
if He was involved. He doesn't make mistakes."

"Some say that You invented the rainbow
to proclaim that You made a mistake,
indicating that You would never do it again."

"Preposterous. Rainbows are an optical effect."

"Do You of did You ever do anything wrong?"

"I can't. I am all Love."

"Did you give too much love, perhaps?"

"Yes, I give near infinite amounts,
but there's nothing wrong with that."

"What was the purpose of having dinosaurs
around for 650 million years,
then extincting them, via asteroids,
plus 95% of all the species?"

"Just playing around; actually, I had nothing to do with it."

"What was the Intelligent Design in this?"

"There wasn't any, for God dos not exist. Can I go now?"

"No, we know that nonexistence trick.
Whose side are you on in football games?"

"I don't take sides or play favorites."

"Then where do humans get all these ideas about You?"

"You know humans—they just make things up."

"Is there a Hell, like maybe in the heart of the sun?"

"No, there is no Hell.
I wouldn't torture my beloved creatures
if I were God. Would you torture a kitten?"

"Some would, but, hey,
it is You who is on trial here, not us.
We only have our human nature
that You have given us and it can often go astray."

"True, as to the human mammal recipe,
plus I am a nice Guy, the nicest ever.
I would not fill your cup to the brim
with temptations and then expect you not to spill it;
I'm a giver, not a taker. I don't make trouble.
Pure love is all giving; there are no strings attached."

"Thanks. Does our free will have to match your will"?

"Heavens no, for that wouldn't be free will, would it?"

"So, there's not even a purgatory,
like somewhere on Venus?"

"Negative."

"How do humans come up with all these things?
They make You out to be some kind
of strict enforcer father figure type."

"That's it; they modeled the family experience."

"Is there a Heaven?"

"Yes."

"Ah-ha, where is it?"

"On Earth. What more could human beings want?"

"Oh, well they want everything
and even think they are special and above all else,
some even above their own kind."

*"Nope, humans are just as organic
as anything else in nature. Anyone can see that."*

"Well, we have imagination."

"Yes, a gift of Nature, but that's all it is."

"Did You publish a book?"

"Yes, but no, for ghost writers wrote one."

"Any movies coming out?"

"No, it would be hard to beat 'The Dark Knight'."

"Were Commandments were ever issued?"

"Love does not command; it frees."

"That's true.
So, You are innocent of all charges
and plead not guilty?"

*"How many times do I have to tell you.
I am Absolute Good."*

"Ever tell a white lie?"

"No way, Jose. I am the Truth."

"Ever peek at a naked person."

*"Of course, people are made that way.
If God didn't want it that way,
they'd be born with clothes or fur.
Some fools even put fig leaves over Eden's artwork."*

"I must confess to You, God,

that I sometimes think of people naked."

*"No sweat, plus that's also a way
to make public speaking easier.
I am naked Myself. It's OK."*

"Ever stick gum somewhere, when no one was looking?"

"No, for I was looking."

"You are a saint!"

*"Higher than that.
I am Perfect, at least before
I got conceited about it."*

"Ah-ha."

"Just joking."

"Did You make Cosmic Jokes,
like, in sexual human anatomy,
putting a toxic waste dump
near a recreation area?"

"God does have a sense of humor."

"How come You didn't give humans everything?"

"If I gave them everything, they'd have no place to put it all."

"A dictionary has 'everything'."

"In a way, plus Wikipedia is good, too."

"How come birth certificates
have expiration dates,
some even sooner than later?"

"They must, otherwise evolution wouldn't work."

"Did some monkey types descend from the trees?"

"Yes, for your DNA matches theirs 98%."

"So, evolution is true, but not You as a Creator?"

*"I keep telling you,
leaving signs all over the earth, you fossil to be."*

"You don't rule or lord Yourself over anyone?"

"Love serves; love does not rule."

"We have witnesses to some of your crimes."

*"No one can witness Me,
besides, they made all that up."*

"Likely story.
Did you choose a tribe and tell Moses
to crush some other tribes?"

*"Those are just ancient Jewish legends,
along with some history."*

"How come Moses didn't ask for directions
when he was lost for 40 years in the desert."

"He's a man; they never ask."

"Ever let someone just make it
through a developing traffic accident?"

*"What! And let some other poor sap
get hurt or die instead?
You don't know Me very well."*

"So, You don't write scripts
for our human soap operas."

"No, for truth is stranger than fiction."

"Why are You invisible?"

"I am a figment. Have faith."

"What's faith?"

"Belief in the invisible unseen unknown."

"You can't get off the hook that easily.

We can still try you in absencia."

"I'm being very cooperative."

"Thanks. Now, Mr. God, Sir,
Did you send a plague of locusts
to harm the welfare of humankind?"

"I wouldn't think of it;
harmful options don't even surface
in my mind for consideration."

"No lightning bolts?"

"That was Mother Nature, not me."

"Well, as you are a self-made Man,
then what stuff did You use to make
Yourself out of, plus all that is?"

"I didn't make all that is;
I only made Myself
out of the fundamental stuff available;
then I accidentally made humankind
from the same stuff,
some debris that I threw out."

"So, you are not at all responsible
for Mother Nature's doings?"

"No, nor did I make the universe, for I am made of it."

"You are not fundamental and absolute?"

"No, for a system of mind and emotion
like Mine or yours requires parts.
I am perfect, however."

"That's still a lofty position."

"I am just fortunate to be as I am;
I never look down on anyone lesser;
My talent is a given;
I can't even really take any credit.
I am just further along
in evolution than you are.

*Cats, too, have reached a kind
of perfection for their form."*

"You evolved beyond the material plane?"

*"Yes, I am pure waves and fields
and thus not seeable.
You all will get there someday, too.
I just helped you all along the path,
with only your best interests at heart."*

"We will all evolve to become Gods, eventually?"

"Certainly."

"You don't interfere in our world on Earth?"

*"No, for then you would miss all the fun.
Knowing everything is not really that great."*

"There would be no surprises."

"Exactly."

"Do you overrule all or part of reality in any way?"

"No, I'm not bossy."

"Do you underlie all or part of reality in any way?"

*"Nope, as I said, I am in this universe
and therefore of this universe;
I am just higher up the food chain."*

"So, in our terms,
you are just a very powerful but loving alien."

*"That I am.
and if any hostile Ones approach me,
I will defend Myself."*

"Thanks, for that may help us too."

"True, but you are all completely free to be and do."

"How come You allow-give this to us?"

"It's the greatest gift that Love can give."

"Thanks, again.
You seem a good Guy,
but we still have a few more questions,
plus, you know, we can't really consider any gifts
that You gave to us when we make our ruling;
I hope you understand,
for we are often approached with bribes."

"Money talks."

"For me it just usually says 'Goodbye'."

*"But when it returns you might say,
'Hey, glad to see you; I've missed you;
where have you been all my life?'"*

"You're a fun Guy.
So, what is all this holy-holy admiration stuff
that humans do in and for your Name?"

"I don't know; it's really weird, isn't it?"

"I thought You knew everything."

*"Well, by staying out of the way,
I choose not to know."*

"What made the stuff
that we and You are made of?"

*"I'm not sure;
I only know everything
from Me onwards;
that stuff could have
appeared in the universe
from somewhere else,
or have been here forever,
or appeared via some kind of possibility;
it is not marked as holy or unholy."*

"Well, that's immaterial, anyway.
Back to our probe."

"I ain't never did anything terrible nohow!"

"Ever do anything wrong at all?"

*"I threw some litter into space
because there was no where else to put it."*

"What litter?"

"Excess atoms that then made your world."

"Well, no harm done."

"Thanks."

"Do angels exist, having wings and all that?"

*"No, not as humans have defined them.
Wings are useless in space; there is no air.
There are more ETs than Me, however."*

"We thought so.
Is there a Bigfoot?"

*"Ha, ha. Those are just hoaxes put forth
by some hicks in the southern US."*

"Isn't 'hick' a bad name?"

*"No, I am just describing an actual fact,
for which the word 'hick' is perfectly descriptive.
I have to use words that you can understand."*

"So, You've never been seen,
and just about everything bad
that was said about You
by humans is false; so, what's left?"

"Not much, just Me as not 'God'."

"But You created us; you helped us along."

*"Well, in a way, but that was quite inadvertent.
You would have formed
somewhere sometime anyway.
Some of my 'trash' formed*

your solar system;
then you evolved.
Your population was down
to less than a thousand once,
and I guess some of my
good vibrations rubbed off on them
as I passed by on my way
to pick up some rare elements on Pluto.
I was building a new house
that could withstand all eternity.
The weather in space is always bad;
it's full of radiation of all sorts."

"Strange weather all over the earth, too."

"There are many hurricanes that began
from a hint of a wisp of a breeze."

"Mr. ET, is there way to tell
the future of the weather?"

"The 2015 Farmer's Almanac just came out."

"So how do we speed up evolution?"

"Takes time,
but you could enhance
your own chemistry,
As I did."

"Sounds dangerous."

"It is; I was a Jekyll and Hyde for a while."

"Ah-ha, that's when
you committed crimes against humanity!"

"No, I was far away,
plus that was 35 billion years ago."

"Oh, but do You have an alibi?"

"No, I was all there was then,
but I have pictures."

"Let's see."

*"I don't have then with me,
but they are very similar to those
taken by the Hubble telescope."*

"You were there among those
trillions of stars and galaxies?"

"Yes, but I was already semitransparent by then."

"It would be like one of those
'Where's Waldo' puzzles."

*"You'll just have to take my Word
if you cannot prove otherwise."*

"What is the purpose of life?"

"To live."

"What is life?"

"You are life."

"Is life and all really just a bunch of
atomic spinning things
of various compositions?"

"That's it."

"Nothing more?"

"There can be no more, for that is all there is."

"Why do we keep hoping for more?"

*"Greed and having no gratitude, but
still, you are a sparkling billion year product,
and so you are quite amazing."*

"We are pretty cool when you think about it."

"That's all it takes to appreciate life."

"Any other universes?"

"Sure, but many did not amount to anything;
however, I am going on vacation
to a good one next week."

"Be sure to send a post card saying
'Wish you were here',
that is, if there is oxygen there."

"Will do.
Lucky for you here that bacteria and plants
came about and made oxygen.
Thenceforth you began as you."

"Yes, a lucky break;
oxygen was a mere waste product
from photosynthesis."

"See, all is as it seems.
No need to invent any supernatural Intent
to blame or thank for anything."

"All is as it did?"

"Yes, that's why it took so long."

"Indeed, a true God type Creator
could have done it instantly,
not even needing 6 days,
or getting tired on the 7th."

"Yes, but the All is an origin, not a Creator.
The ground-state was always around,
and so there was no creation, and no Creator."

"Yikes, then what should we do?"

"Just be."

"OK, good advice, but,
if we ever find that there was a Culprit Creator
Who committed some of the very crimes
that His commandments spoke against,
like murder, destruction, or hatred,
then He is really going to be toast."

"As He should be,

*for those acts would
have been unconscionable,
especially for Someone
of that high stature."*

"Thanks for your testimony.
We'll call it the Third Testament.
Your judgment day is near at hand.
I'm calling a one hour recess."

"All please rise."

"The court finds You not guilty on all counts,
due to lack of evidence, plus Your good nature."

*"Evidence for those like Me
is not even conceivable."*

"True. Thank you everyone.
Please bring in the next case."

Austin walks in.

"Austin, did you leave the toilet seat up in a household
where there were females present?"

"Well, maybe, yes I did, but..."

"100 years of hard labor in Siberia."

THE ANSWER TO ALL

My office is in the boiler room, and here
In this semisecret chamber from which I write
Are many fine treasures:

I have the one and only jewel-encrusted edition
Of the 'Great Omar' (Rubaiyat) that I fished up from
The Titanic lying on the floor of the North Atlantic.

Here, as well, Aristotle's 'lost' book,
'Beyond Metaphysics'. and, too,
I have some nuggets of gold found
In the original Garden of Eden that I located
In the heart of the Amazon Jungle,
Wherein lie massive fields of Lady's Slippers
And all of the flowers of paradise.

I reached up—and put the apple back on the tree.

And the Celtic Chronicles, I have, that I found
In an iron box beneath Glastonbury Abbey,
Telling all of the tales from the Dark Ages,
And, from the tomb of the Holy Sepulcher—
The Holy Grail itself.

Here, as well, a sliver of the true cross,
A small vial containing a drop of the Virgin's milk,
A pebble, from a moon rock, given to me
By a polymath who works for the President,
A smart thinking and talking cricket named 'Crick',
The spear tip that pierced the side of the Saviour,
A few molecules of immortal air
From a sealed pyramid chamber in Egypt,
Some secret papers retrieved from the shaft
Of the bottomless CIA trash pit
Of "things that never happened",
A thriving rose bush, just outside the window,
That was begun from Omar Khayyàm's rose garden,
'Flamberge'—Prince Valiant's 'Singing sword'
(Twin to 'Excalibur'),
Thomas Jefferson's briefcase,
An original and intact Ming dynasty vase,
The third [missing] tablet of the 15 Commandments,
And the solution to gravity,

As it is a means and a reason
For quantum collapse from superposition,
As well as a tennis ball with my initials
Marked on it in a yin-yang style.

Yet, all of these treasures pale in comparison
To reality's truth unveiled, but no one cares about that.

I also have the 'treasure' of a preliminary,
But solid indication of the Higgs particle's existence,
Which Lisa Randall was nice enough to give me
From the LHC's latest analysis.

I am now holding part of a brick that came from
Nero's very recently discovered revolving banquet hall
That kept pace with the turn of the Earth.
I am about to ponder the existence of this brick,
But that would probably be too disruptive to my life,
So I'm going out to date some old fossil instead...

I'm back—and she is very young at heart
And quite exciting, so we are trying to tone it down
By smoking some pot and pondering the brick.
Just kidding.

Actually, I'm thinking of the Library of Congress,
For I heard that it has five hundred miles of stacks.
It began anew, after burning by the British,
When Thomas Jefferson donated his personal library.
I found his personal diary
In the lining of his brief case.
It said the founding fathers wanted to retain a Deity
To save the new nation from the religious
Superstitions associated with a Theity.

I hold in my hand a bone from
Early sapiens or of proto-man.
He is not gone, though,
But lives on in your heart and mine,
As in him lived all those before
In which the universe itself came to life. Amen.

Yet, all of these treasures pale in comparison
To reality's truth unveiled...

Why Anything?

The human condition is such
That it often just prematurely halts at a word,
Such as 'God', for the believers,
Or 'matter' or 'forever', for anyone.

The Cosmos or its basis, meaning All,
Not just our locality or universe,
Must be eternal, or it wouldn't be every-when,
As well as infinite, or it wouldn't be everywhere,
And so the prime and causeless mover
Must have these attributes,
Requiring nothing else but itself.

Nor can the ultimate basis be a complex composite,
For these are not fundamental, but come later.
('God' is out, too.)
The basis must be the simplest elemental state.

As for matter, it has many particulars,
Such as its total amount and its individual properties
Of spin, charge, form, size, mass, location,
Matter vs. antimatter state, and other specifics,
Or limitations, such as
That there are only two stable matter particles,
The electron/positron and the proton/antiproton,
And only one stable energy particle, the photon.
(Neutrons decay.)

We cannot just stop at the word 'matter'
And just say that it is what what was around forever,
For one simply cannot have an eternal something
Already made and defined in all of its particulars
Without it ever having been made and defined
In the first place that never was.
Impossible.

So, where does this leave us?
We are fine, for there is/was literally nothing
To make the original stuff of, anyway,
And there is no way around this fact;
So, 'nothing' must be the answer,
It also being the simplest state,
One that is necessarily perfectly unstable,
For it cannot be at all or stay as such.

So, the vacuum fluctuates,
Making the vacuum only a 'vacuum'.
Movement is natural, not stillness.

Existence is a positive/negative distribution
Of nonexistence.
'Nothing' is the only candidate for the prime mover.

Welcome to zero-sum physics;
(And 'nothing' is exactly the opposite of 'God'.)

Look about; there are particles
Of opposite polarity of charge
And matter/antimatter states;
The weak force opposes the strong force;
The positive kinetic energy of stuff is canceled
By the negative potential energy of gravity, etc.,
For an equation of a zero balance
Had to replace the cause and effect
That could not have gone on forever beneath.

It is the opposite polarity of charge
That nullifies all of existence in the overview,
But not in actuality, for nothing cannot be.

Zero-sum physics perhaps started here:

Einstein as a near traffic fatality...

George Gamow told in his book, 'My World Line',
How he was conversing with Albert Einstein
While walking through Princeton in the 1940s.
Gamow casually mentioned that one of his colleagues
[Pascual Jordan] had pointed out to him that according
To Einstein's equations a star could be created
Out of nothing at all, because [at point zero]
Its negative gravitational energy [mass defect]
Precisely cancels out [is equal to]
Its positive mass energy [rest mass].

"Einstein stopped in his tracks," says Gamow,
"And, since we were crossing a street,
Several cars had to stop to avoid running us down".

Now that we know of this zero-balance requirement,

We might use it as a reason
For the necessity of conservation laws.

What about the word 'eternal' or 'forever'?
We need go on to the implications,
For forever systems are their own precursors.
No first matter making light;
No first light making matter.
No first anything.

How? Opposite pair production, perhaps,
Or that infinity times zero = one;
Take your pick.

Boundless space, overall electric neutrality,
And conservation of charge, momentum, and energy
Leads inexorably to nothingness, really.

The zero-equation is the reason
The universe is the way it is,
The reason why the universe
Must be the way it is,
And the reason why it is.

It is the perfect zero-sum equation.

Zero and infinity, the smallest and the largest,
Both lead to nonexistence,
And so our finite existence cannot be there,
But must be at its midpoint.

Zero and infinity lead to many
Of the same problems in algebra and cosmology.
They are the same thing: nonexistence.

The deathly spiral of paradox ever follows
The carving of wishes into the stone hollows
Of dogma forever blocked from the allowables.

The believing dance grinds to the elemental
Of that Being who can never be fundamental.

All such tales of original stuff made of love
End where there's nothing to make it of.

CRYSTAL MEMORIES
The Seasonings:
From Spring through Winter.

From her hilltop cabin of logs, Sin-thea
Recalls the ice, once in her veins, the freeze;
Lo, the Canadian lilacs have bloomed, at last.

Nature springs from winter's tomb,
The bloom already in the seed,
The trees within the acorns.

Crystal fragments remain, sharp memories
Of the ventures in which she shattered not.

Surging sprigs sprout from the soil;
Spring showers make the summer flower.

The seasons arrive at her door, in turn,
For all things come round to those who observe.

Summer wakes from spring's dying kiss,
Blooming when the rose does,
Sunning after the spring's running.

She could never be too warm,
For she'd endured the frost.
The kaleidoscope revolves: life's cycle.

Summer reigns upon the land,
Eventually fading in the night.

Life's second bloom shines upon middle age,
Colors her mind, its rainbows shimmering.

Autumn falls as summer leaves,
Harvesting its sum of days,
Seconding the rose of spring.

The hearthstone fire glows heartily, with her self,
As she stokes the flames of the wondering soul.

The smile meets the tear;
Fall's embers last through December.

It snows atop the trees, ne'er falling in,

Entombing the spring that waits for the miracle.

Ice winds stalk the weed flowers,
The ghosts frosting the dead stalks,
Snow crystals barring all that grows.

She's in the cabin safe, snug, warm, and whole.

Winter is life cooled over;
Melting snows feed spring waters.

SEGNO (SIGN) # 5

"There's the aether", replied the other, apace,
"The 5th state, one that pervades all of space,
Yet there are no signposts of it up ahead,
Or anywhere, since it's in every stead.

"We regard it as the stuff of which Gods are made,
That lively spirit of elixir that their nature bade,
For, just as all mortal creatures inhale the air,
So do immortal and divine natures inhale the aether."

"The intimation is the mark of their manifestation,
A demonstration and a token of the evidencention—
The aetheric and heavenly sign of things to come,
Both the portent of the miracle and its omen.

"It is of the warning and the notice let,
Presaging both the promise and the threat.
Of this sign the aether follows, the gesture beckons;
'Tis the signal, the wave and gesticulation reckoned.

"We can read the writing in the sky, the marquee,
Daubed with symbols marking the cipher free,
With characters, figures, and hieroglyphs of time,
The ideogram of the rune, the emblem of the Divine."

THE IMPOSSIBLE RECIPE

Explaining the Cosmos is as easy as pie:
It's an endless extravagance beyond the sky,
Which shows that matter's very readily made—
Underlying energy raising the shades.

This All sounds rather like an ultimate free lunch,
For the basis is already made, with no punch,
It ever being around, as is, never a 'was'—
Everywhere, in great abundance quite unheard of.

There's even more of it than can be imagined—
Of lavish big spenders, there in amounts unbounded:
Bubbles of universes within pockets more,
Across all the times and spaces beyond our shore!

What is the birthing source of this tremendous weight?
There is nothing from which to make the causeless cake!
Its nature is undirected, uncooked, unbaked?
There can't be a choice to that ne'er born and awaked!

There can't be turtles on turtles all the way down;
The buck has to stop somewhere in this town.
'Nothing' is unproductive—can't even be meant;
All ever needed <u>is</u>, with nothing on it spent!

Yes, none from nothing, yet something is here, true;
But, really, you can't have your cake and Edith, too!
And yet I've still all of my wedding cake, I do—
It's just changed form; what ever IS can never go.

Since there's no point at which to impart direction
The essence would have no limited, specific,
Certain, designed, created, crafted, thought out meaning!
Thus the Great IS is anything and everything!

This All is as useless as Babel's Library
Of all possible books in all variety!
Yes, and even in our own small aisle we see
Any and every manner of diversity.

The information content of Everything
Would be the same as that of Nothing!
Zero. The bake's ingredients vary widely,
And so express themselves accordingly.

What's Everything, detailed? Length, width, depth, 4D—
Your world-line; 5th, all your probable futures;
6th, jump to any; 7th, all Big Bang starts to ends;
8th, all universes' lines; 9th, jump to any;
10th, the IS of all possible realities.

Your elucidation is quite a piece of cake!
Yo, it exceeds, as well, and so it takes the cake.
Everything ever must be, because 'nothing' can't?
Yes, it's that existence has no opposite, Kant!

So, we're here at the mouth of the horn of plenty,
For a free breakfast, lunch, and a dinner party;
Yet many starving are fed up with being unfed.
Alas, for now I have to say, Let Them Eat Cake!

NEVER MIND

The first cause could not be of Mind Aware,
For a complex composite's parts must precede.
Yet think no more of things before some things,
But of 'before' existence and physical laws.

EXTRA SENSORY

Our instruments detect what our senses cannot,
Of the whole electromagnetic spectrum,
Of odours and molecules beyond smell and sight,
Of stars far away and even way back in time,
All because scientists exalt in mystery.

Human introspection and sensation, alone,
Without being informed by the science available,
Is captive to the tales of its second story,
Not knowing the neurological first storey.

So it goes on to declare wishes and beliefs
Truth, deepening the wiring upon each visit.
And they might then layer more dogma thereupon,
Till they've a far ranging scheme for life's wonderland.

Things that haven't been established can't be addressed,
For they're 'invisible', such as evil spirits,
So really, a belief in the stated unknown,
'Faith' as defined, can't even be known, much less shown.

Thus to feelings, senses, desires, and sensations,
And claims, we can't trust, true, even their wishing point.
Mysteries shrink away, at an alarming pace,
Now-a-times, and it's hard to keep up with the race.

As for humans, true, we, and our matter that's bright,
Seem to be an afterthought of the Cosmic scheme:
We glow-surf on informational waves of light,
A tiny minority in the grand regime.

Science discovers the truth deep within everywhere;
Religion just makes for begged and bigger questions;
Philosophers just sit around in their soft chairs;
Evolution explains how we mammals got somewheres.

PHANTASMAGORIA

Each morning as I soft awake and slowly stir,
Remembering, I can just barely recognize her.

Even as I rub my eyes, in their own reflection,
She becomes but a shadowy recollection,
Although a most pleasant and enlivening one,
As my day begins anew with star set and sun.

In night's stillness, fancies wend their utopian way,
Unimpeded in their stay, safe from raucous day—
Where they quick become fugitive and transitory—
Evaporating, melting away, and disappearing.

I didn't even catch her name clearly,
Yet I know that I love her dearly.
How could I not, for I've created her
In the most perfect, loving image.

She is a dream, ephemeral and transient.

As the morning wears on, she's still with me, in part,
Although but the faintest glimmer of being in my heart,
A mere shadow of evanescence, but for love's sake,
Of the fading impression I felt at daybreak.

As the day grows onto noon in its brightness
My remembrance of her dims into vagueness.

By late afternoon, I lament, she's but a wisp
Of near nothingness, further vanishing away;
Yet I still can feel a temporary presence
Of her joyous but fleeting fulfillment present,
As if she had somehow snuggled into my being.

Oh for the deep clarity of the silent night,
When our being mirrors the stars and the moon light.
She has merged into me, though short-lived by day,
Impermanent, but who is she; what wrote her way?

Well, she seems to be every woman I've known,
Yet none in particular, plus ones that I've grown.

Even now I'm having trouble rebuilding her.
It's so hazy now; if only I could recall.

Somehow, I must see her again distinctly,
In the still night's clear sight, and more importantly,
Remember the lovely vision perfectly,
But how can I become alert, conscious, awake,
And clearheaded sober of thought in a dream state?

Several nights flew by; I didn't dream of her;
But then, just out of the silver moonlight, alight,
On one rare intoxicatingly drowsy night,
I saw her again, and I lived and loved long with her
As if tomorrow never was, would never be;
However, all too soon the dreaded morrow broke,
And therein the light she waned, lost to me again.

Although she was so vivid and intense, in sense,
She faded to a familiar evanescence,
But I managed to write down her fleeing description,
And by that evening ringing chime her depiction
Was all I had left, a faint but fine impression.

Though her representation was wont to wither fast,
I was now able to quick resurrect her past,
Using my hasty and foggy written description,
Even though it was made in an all too sleepy shade.

For awhile I could capture her visage as such,
But again her dear image faded all too soon.

Many phantasms ran through my mind the next night,
Overturning into ghostly visions of fright,
As all nonsensical hallucinations
Of the most illogical character rations;
That is, I was dreaming fantastic nightmarish dreams;
But, my dream girl was not in any of the scenes.

If only I could bring some order and fair sense
Into the senseless, noisy, and mosaic mess
Of my random and wandering thoughts in darkness.

Several notions even waited in the wing's herd,
For their appearance on the stage of the absurd,
And the mass of those scripts soon tumbled and stumbled
Across the scenes on the stage, had their moment, mumbled,
Then passed into oblivion, never to return.

I saw them all pass by, as the lone spectator,
And since a good part of my mind was out of life,
Found nothing unusual in the chaotic strife,
And therefore had believed it all to be quite real.

The weeks crawled by, and I don't believe I saw her,
But if I did I must have forgotten the bond;
However, our love continued to live on an on,
As that romantic idea painted in me fond.

I'd an inspiration for a presentation:
If she wasn't going to show up on a station,
Why couldn't iI conjure her up through imagination!

It took many days of meditative practice:
I went to bed relaxed, after a nice warm bath,
And thus easily discarded all the day's chaff.
I then reviewed and read out the script in my mind,
Going over it and over it many times.

Control your dreams' static going automatic;
Dreamscapes aren't reality, though they seem to be,
You can do anything that you wish in your dream;
You can guide and control it. It's only a dream.

Tell yourself therein that it is only a dream.
Grasp the idea's core and then become lucid.
You can do anything, go anywhere, at will,
See or be anyone, have anything, bidden,
If you can only realize that it is a dream
And then direct the dream's contents accordingly.

I repeated the words while I tried to picture there
The most utter and complete blackness of nothingness...
And in there I etched the words rehearsed above,
So that they would remain, floating, the only link,
As a message to myself after I soft slept—
To my normally unbelieving dreaming self,
The drowsy mind that seldom questions ill logic,
The mind that interprets dreams so literally,
Because they all do seem so real, internally,
Which is because the model employed in dreams
Is the same model that's used when we are awake!

I looked forward to the night, with anticipation,
Wondering if dream images were really sharp,

Clear, and distinct, or vague, as in their remembrance.

Soon I would know the answer. *It's only a dream...*

These were the last echoing words I awake heard
Before drifting off into that faux nether world
In which I hoped to script, direct, produce, and star
In any narrative that I could then dream up.

And there, in my reverie, the inscribed thought
That I was dreaming did indeed occur to me.
What a revelation it was! A realization!

Still, it seemed to be so far-fetched, so amazing,
That I hesitated to believe at the time.

Why didn't I fully believe it, virtually?
Because everything in my dream was very clear,
So sharp and colorful, a just perfect image
Of reality itself in three dimensions—
An exact match to our actuality itself,
A genuine reconstruction of reality!

The next night I was again haunted by the strange thought
That *I was dreaming.* I still wasn't convinced,
But I took on some cautious control, anyway,
So that I could try a harmless experiment.

I went down the stairs to the kitchen in my dream
Opened up the fridge, and poured some milk on the floor,
Much as it pained me to do so, though cleanable.

As soon as I awoke the next morning early,
I rushed downstairs, noting the kitchen floor was clean!

This gave me confidence. I was making progress
Toward control. I was learning to detect the dream state.

The following night I dreamt again, and took control,
Realizing that it was all a fabrication.

This time I rearranged the wonderful items
That were on my bedside table, and when I woke up,
They were still in their original positions.

I was getting close; I was starting to believe.

I had to be careful, a maybe and perhaps,
Before doing any crazy things in my dreams,
For one must be surely absolutely convinced,
Beyond uncertainty, that a dream is a dream,
Lest one fall into harm or become inhibited
Out of fear of breaking the law or passing away.

The next night in my dream I wondered again
If I was dreaming whilst flying down the street,
About twenty feet in the air—now a big clue.

The logical part of my brain fully "awoke",
And said to me,
"You are flying down the street twenty feet off the ground;
This is impossible; therefore this must be a dream!"

I was thus thoroughly and utterly convinced
To the core of my being that I was dreaming.
Now I could begin some serious dream research,
Living it and observing it at the same time.

Instead of transporting to some paradise,
I first wanted to inspect all my surroundings—
To minutely analyze the dream's images.
So I made a conscious and definite effort
To look directly at everything in the scene.

As I flew through my neighborhood, out of body,
I looked closely at each house, yard, plant, and construct,
And saw that every part was perfectly in place:
Every shingle and nail, each blade of grass distinct,
Every leaf and branch vivid, all as ought to be;
In fact, every fine detail, including color
And odour, identical to that of real life,
Indistinguishable from it! What a discovery!

I flew high, low, and far, in this second world.
The reconstruction of my street was finely perfect;
No wonder that dreams seem real; They practically are.

Of course, dreams may seem hazy, after awakening,
But that's only because the recollection itself
Grows hazy over time; however, I have found that
If you write your dreams down just upon awakening
You will find later, upon reading about them,

That they can be thus recalled, remaining vivid.

So thusly, after many weeks of such patience,
Discipline, meanderings, and the use of dream notes,
I was able to do what I wished in my dreams:
I toured; I ate delicious food, met with people.

I soon formed plays, movies, and otherworldly scales
In which each player performed in their character,
Many of which were unlike my own character;
Yet all their performances must have come from my brain,
And fine scene after created scene rolled by
In 3-D Cinemascope and Technicolor.

I could now near do anything that a god could do;
However, it was time to find the phantasma
Who'd initiated my quest in the first place.

She came easily into my nightly vision;
I loved her; for she was made wondrous just for me.
She was my heart's ideal molded into being.

Why should I ever wake? Why indeed. Life is harsh,
And I had just stumbled right into Heaven on Earth.

Well, one must wake to live, to make one's dreams come true,
And to gain input for further dreams, which in turn
Give even more desired life upon awakening.

We all have to sleep, and must do so every night.
Why waste it? It's paradise; It's the perfect world,
One in which no debts are owed, where power awaits,
Where you can have a second life there on offer.

...

She awoke that morning from a dream, seeming new
And refreshed, with that free and wonderful feeling
That lies at the heart of life's exhilarating glory;

But soon the old waves of stifling reality
Swept on toward her, like an approaching sickness,
Smothering her in the dread of another hopeless day,
Amidst the ruins of anxiety and depression.

She was like a doomed ship drifting in the storm's aftermath,

Under a cold moon pale and wan, her sails tattered
And torn before the relentless wind of existence.

Her dream had seemed so real, so life like, but it too,
Had wilted in the heat, like a flourishing flower
That had lost its valuable gleam of morning dew.

But the firm hull must drive on, mustn't it, she thought,
Though the mast be broken... No! No more! It's over.
Today I'll end it all; tonight I'll end my life!

She spent some time planning—the finis and the end.
Yes, she would scuttle her ship—her car, at the cliffside,
And sink within it to the bottom of the sea,
A river, really, and drown, with a sigh and a groan,
Devoured by forces too large to fight against.

She drove her car towards the precipice near the bridge.
She drove faster, faster. The waters called to her;
Their cool and refreshing depths invited her in.

"Come to me," some deathly voice whispered in her ear,
"Come to me and find everlasting bliss and peace.
Come and sleep with me in the endless, boundless night;
Let me cover you with my ebon wings in darkness,
For it is eternal, infinite, and complete."

"No, no, not thee!" she cried aloud.
"I can't go with thee, not with evil!"

She drove her car to the edge of the cliff,
Stopping short, now drinking in and dear savoring
The blue-green world reflected on the river surface.

This sort of sparkling day was not the kind of day
On which she could end it all, throwing it away.

As she looked deeper and deeper into the scene,
She drifted into a dream-world of her own making,
A fairy realm in which her ideals could live on,
Untainted by pains of a mediocre world.

A voice called her. Apparitions danced in her head;
Mythical fantasy worlds, mirages, and legends
Beckoned to her, seemingly from all directions.

An inner voice called to her, the old summoning,
The sweet voice of someone like her who she could love.

She'd often retreated to this storybook world,
But now she would take it much further: plunge into it,
Live within its splendor, in all her being, therein,
Residing mostly in that wonder—before all else.

This dreamland would find her saved within its refuge.

The fairylande called to her daily, there awaiting;
It was and would always be the realization
Of many of the imagined perfections
That she had always brought into her wishing mind,
When the real world had often failed expectations.

She freed her mind from its real life shackles,
And thus began to daydream more freely.

"I'll breath life into you, my little voice,"
She said to herself.

As the noise of consciousness slowly faded away,
Her imaginary world came into its focus.
She could now paint it with the colors of her dreams,
Creating life much closer to the heart's desire.

She felt like a goddess, powered by imagination,
Being able to create life at will in her dreams.
This is when she thus so inspired created him.
This is when she brought him to life—from her essence.

His existence was his own to have, free of strings,
And so he knew naught of her as his creator,
Just that he was in a beautiful, perfect world.

She had built him in her soul's best image;
She had molded him from her heart's wishes;
She fell in love with him, for she could do no other.

"Come into my dreams,"
She would say to conjure him up;
"Come into my dreams,
And then by day I shall be well again",
For she was using lines
From the romantic poets she'd read.

He was a good and decent human being,
For how could he be otherwise,
With her ideals brought to life in him.
He gave fully of himself in life and love,
Always placing his partner's happiness
And fulfillment above his own.

Their relationship was driven by love alone,
And they celebrated it often in her dreams.

She had, at last, found the love
That the real world had so often denied her,
For she had created a new and better reality.

He did feel a bit of sadness at times, too,
For she could not totally submerge
That part of herself, but it was subdued in him,
And so the sadness was used as necessary
To enhance the beauty of their love
With its sheer contrast and brightness.

She gave all that she had to him,
Watching over him and loving him deeply,
Utterly, and completely.

Nothing could hurt him in this special world.
He was impervious to pain, cold, fire, and sickness.

Once he was fatally shot in a war, but he didn't die,
Because it was from her spirit
That he drew his life's principle,
And of course she had willed him to live on.

Another time, he was hit by lightning,
But as we have seen, a dream can never die,
And so it was that he arose alive and well
From the smoldering embers.

He never got sick and seldom had a headache.
Everyone should have the best in life,
She said to herself,
And in my world there can be no suffering.

Each night he would come, saying,
"*I arise from dreams of thee.*"

"Kiss me, my dearest phantasm," she'd whisper,
"And hold me ever dear; shelter me
From the evils and the melancholy
Of the torturous world;
Show me the true meaning of love
That the real world has forgotten!
Come into my dreams,
And then by day I shall be well again."

Knowing not that he was her dream image,
He never doubted his own existence and happiness.
When she didn't think of him or when she slept,
He disappeared, temporarily,
Until she awoke or thought of him again.

So when she slept or daydreamed, he existed,
And when she was awake and not daydreaming,
Then he slipped into that oblivion
Which he only knew as sleep and quiet slumber,
The gift of Death's kinder brother.
He was the day to her night.

He arose from her dreams of him,
Much like the mountain rises
From the depths of the valley.

Without her, he could not be;
Without him, she could not be.
The circle was now complete;
The link was closed.
They had become two locked boxes,
Each of which contained the others key.

That he only existed as a dream in her mind
Took nothing away from their relationship,
For their love was true,
And the feelings were felt as deeply
As in the tangible world

Ultimately, it is what we feel that matters,
Not the source that causes the feeling,
For all sensation comes from within.

He did wonder, sometimes,
About just how good and lucky his life was,

About his having almost super powers at times;
But, he concluded only that he led a charmed life
That had stemmed from an inner happiness
That constantly poured forth visions
In positive creative images
That ever bred good fortune.

Indeed he had, for she had given him that power—
A power that had come from somewhere within her.

He was her twin, yet also her opposite,
For somehow she had given him
An enthusiasm for life
Which she didn't seem to fully have herself.

He was a reflection of her image,
In which his outward vision
Mirrored her inward hope.

Consequently, he blossomed with creativity
In art, music, and writing,
As she continued to maintain him,
As both his protector and his inspiration,
Although, as we have seen,
He certainly did seem have 'free' will,
For he knew not the source of his creation
Nor of the tendencies that were placed into him.

So they lived and loved together,
Allied and alloyed in the soft metallic night,
Blending into the golden oneness
That love had always promised,
But had never before delivered.

He was born with the inclination of goodness—
So she never had to possess or demand from him.

Life blossomed now,
And some of this exuberance did indeed surface
And show itself back in her real world,
But in the end she still found her waking life
To be the cold harsh reality that it had always been.
So she called him back to her dreams,
Again and again.

Here they were free to love and live fully,

Their chemistry sending out invitations of love
Which were soft, sweet, and smiling on the rising air,
A fountain's spray of liquid love, mystified,
Filling the scene with the vaporous perfume
Of its well-being everywhere:

They were up, warm,
And floating on the clouds of dreams.
Their passions smoldered like incense,
And burned like the candle's flame;
They consumed each other often,
Yet continued to have endless love to give,
Their passions always seeming to reach new levels,
Then expanding even more, building, ever building.

She had to attend to events back in the true world,
But it really wasn't so bad there anymore,
Because she knew that she had something
To look forward to in her dreams.

So she went happily through the motions,
Feeling better and better as the days went by,
But always looking forward to the chance
To dream him up again,
When she would say softly to herself:

Come to me in my dreams, and then
By day I shall be well again!
For so the night will more than pay
The hopeless longing of the day.

Come, as thou cam'st a thousand times,
A messenger from radiant climes,
And smile on thy new world, and be
As kind to others as to me!

Or, as thou never cam'st in sooth,
Come now, and let me dream it truth,
And part my hair, and kiss my brow,
And say, 'My love! Why sufferest thou?'

Come to me in my dreams, and then
By day I shall be well again!
For so the night will more than pay
The hopeless longing of the day.
(Matthew Arnold)

She again faded off into dreamland...
And there he was.

Just the sight of him
Would bring the world to a stop,
For she could only concentrate on him.
When she looked at him,
The birds' song fainted on the moving air,
The night breezes stopped in their motion,
And the moon's radiance shone no more,
For her heart had welled up within
And had merged with his own.

She felt herself being drawn deeper
Into this dream of love,
In which there was only one overwhelming
And all consuming feeling: glory, peace and unity.

But then,
During one rainy night back in her real world,
When she was driving in a storm,
Along the cliff road around a curve,
Where she had once contemplated suicide,
Her car skidded,
And flew off the side of the water-slicked road,
Falling three thousand feet below,
And crashed hard and straight into the rock,
And then exploded in a fiery wreck.

The flames licked at her for hours, but she felt no heat.
All her bones should have been crushed in the fall,
But they weren't. She did not even bleed.
There was no pain.
She arose from the car's wreck unharmed,
And walked away.

It was then that she realized that she too
Was a character in someone's dream...

...

... She did not even bleed.

She was a figment
Of someone else's imagination.

"Who dreamest me?"
She cried to the sky.
"Reveal thyself! Who art thou?
Who art thou that won't even let me die!"

The heavens remained dumb,
So she climbed back up towards the road.

Back at the top she again cried,
"Who hast made me? Who?—
Thy image is tainted,
Thy DNA is corrupted!"

Visions of angels appeared in the sky.
"You have a question for us?" they asked.

"Yes, what sort of Being made me
To suffer and toil in this sad world?"

"It's a lovely and beautiful world,"
Said the angels in a chorus.

"OK," she said, "I'll play your game,
Shelleyesque. Tell me now,
Who made this varied and sensual world
Of charm and grace and color?
Who gave me intellectual beauty,
And those rare but beautiful waves of emotions
Which I have known and enjoyed
For their breathtaking meaning and depth?"

"A good and loving spirit," they said.
"That's our usual answer."

"And who gave me freedom
To love and live and grow,
Flowering free and fragile,
Though beautiful, but then withering,
Faded and forlorn in old age,
As an evanescent dream?"

"It was the Creator of all life."

"And who gave me sadness?"

"He did," they answered.

"And who gave the world hunger, pain, misfortune,
Sickness, death, worry, and unbearable calamity
Which drags us suffering to the grave?"

"He reigns," they said.

"Give me his name!" she asked. "Who is he
That does not even grant me peace in the grave?—
For Hell awaits me there as a further torture?"

"He rules," the angels replied.

"His name! I ask but his name—
The name of one so cruel!
Who is the one that would create man
As a precious vessel, quite imperfect,
And then destroy this lovely creation
By sickness and death, in rage?"

"He is the One," they said.

"Name him and let him be known
For his vengeful name,
For in my own fine dreams of a man
I allowed no sickness,
No pain—all was love and beauty!
I out-think this so-called master.
Who is he that is the source
Of my everlasting pain?"

"He does not exist,"
The angels finally said,
"Nor does the Devil, nor do we;
All is simply virtually as it is
And so it ever shall be.

"There is no 'why', for that would be 'purpose,'
And beings came later, with their 'whys'.
There is only 'how', which is as causeless.

"It's the way that the universe happens to work.
Therefore, all is right with the world.
We angels are simply manifestations
Of your own thoughts.

All that is truly real comes from within;
Nothing comes from without."

"There is no creative deity?" she asked.

"There is none;
There is only the unconscious luck towards life,
Which is part and parcel of the universe,
Co-eternal with it and embodied in it
As the principle of movement in all things.
It is the connectedness of everything,
And exists far below the level of atoms."

She didn't know whether she was relieved or angry,
Not having anyone to blame for the state of the world.

"But whose dream am I,"
She wondered aloud.
"Who saved me from death?"

Another voice now replied—
The familiar voice of the man of her dreams.
"It is I who made thee, my beloved," he said.
"I dreamt of thee.
You are the dream of my dreams;
You are my ideal,
For your love is so innocent and free!"

"No," she said, "It cannot be,
For it was I who made thee in my dreams."

"Yes, as well," he said,
"But my image was already in you, was it not?
Who put it there?
It was from that image
That you gave birth to mine—
But the real story is more like
That we have somehow made each other.
I may be the day to your night,
But you are the reverse to me
When I dream of you.
I am your opposite twin.
Neither of us can exist without the other."

"I believe it," she said,
"Although there seems to be no initial cause.

Very strange though."

"I see and dream of you, my dream woman,
Each night," he whispered.

"We are indeed two souls,
Each of which opens the other,"
She replied.

"Yes, it is I who made you as you made me,
From all that was already inside us.
As your twin spirit I arose,
Given life only by your dreams.
Oh please, let me live, for now I sustain you—
I protect you and love you
As you do the same for me.
And now that I love you and want you,
I need you."

"If one of us dies," she said,
"Then the other would perish also?"

"The valley cannot exist without the mountain.
There can be no day without the night;
There can be no beauty without sadness,
No yin without the yang.
We are twin-opposites,
As alike as dawn and dusk in our aspects,
Reflections, as it were of each other's image—
Visions which truly exist in the mind,
For all is real in the mind."

"Day gives birth to night
And then night gives birth to day.
That is us and that is the cycle which created us,
Within which scheme it was not necessary
For either part to come first,
As with the chicken and the egg."

"But we live neither here nor there.
Does it matter?
Now that we know that we're just dream images
How can we really live and love?"

"We can neither fully live
Nor completely die where we are.

What is deathless is also lifeless,
Although it is still a beautiful work of art,
Such as the ideals that we see in a painting.
I can be as real as you wish me to be,
As can you to me."

"Some say it's crazy to try and live a dream."

"Some say it's crazy not to!"

"Join my real world," she said,
"And I will join yours as well."

"But your day is my night and vice versa.
How can we meet?"

"We'll meet at twilight dawn or dusk,
The only time that night and day can touch."

"I shall come," he said,
Leaving his dreamland forever
And joining hers as her real life love.

She greeted the man of her ideals,
Saying to him,
"I have wished you into being.
My thoughts of you have colored my actions
And have led me to find you in the real world;
It was a self-fulfilling prophecy,
An example of positive creative imagery."

"It was indeed," he answered.
"Although here I shall at last know
True sadness and death.
But, also,
I will experience higher levels of beauty."

She said, no longer anxious or depressed,
"When you're open to beauty,
Then you become vulnerable to sadness.
What I have finally learned, the hard way,
Is that they are inseparable in life."

"Some people lead lives in which
They are fat, dumb, and fairly content."

"Yes, they don't live much, but then again,
They don't suffer much either.
They're immune to both beauty and sadness."

"It's like when you're not with me.
There is pain when I miss you,
But for me, if I had no one to miss,
Then the pain would be greater."

The new light of morning shone
In that blessed mood
That attends to the quiet intermingling
Of day and night
In the dawn's misty twilight.

She came to him during morning twilight;
He came to her at evening twilight.

In between, they dreamt of each other.
Each day forward was born in quiet innocence
As their human hearts tenderly touched—
Open, vulnerable, and exposed,
Yet fully alive and beating.

Days turned into weeks,
As they grew close together in the soft glow
That was neither night nor day,
But was somewhere in-between,
In that nether world of half-light dawn or dusk.

The morning brimmed with the freshness of life,
Its beauty spreading far and wide
Into every root and tendril.

Life took wing from the cocoon,
As caterpillars having magically transformed
Into beautiful butterflies.

Weeks turned into months.
It was a dream within a dream within a dream.
Faint images from dim shadows
Flickered and grew brighter.
High noon came and showered its brightness
Into life's every chamber.

Now that they had felt the glory of reality,

They would seek it always.

From the months a life was made.

The afternoon sparkled
And spread its gold to every living thing.

Years of contentment rolled by.

The soft light of evening shone again,
As always, in that sacred mood
That attends the quiet intermingling
Of day and night in the twilight of dusk.

He came, as usual, saying:

I arise from dreams of thee
In the first sweet sleep of night,
When the winds are breathing low,
And the stars are shining bright.

I arise from dreams of thee,
And a spirit in my feet
Has led me—who knows how?—
To thy chamber window sweet!

The wandering airs they faint
On the dark, the silent stream,—
The champak odours fail
Like sweet thoughts in a dream,

The nightingale's complaint,
It dies upon her heart,
As I must die on thine,
O, beloved as thou art!

O, lift me from the grass!
I die, I faint, I fail!
Let thy love in kisses rain
On my lips and eyelids pale.

My cheek is cold and white, alas!
My heart beats out loud and fast
Oh! press it close to thine again,
Where it will break at last!
(Shelley)

He awoke that morning from a dream,
Filled with dread, dripping with sweat,
Wondering whether he had gone
To Heaven or to Hell,
Not knowing if he was truly awake
Or still in the midst of a nightmare;
But soon a calming wave of peace
And quiet swept over him,
As he turned and saw that his dream lady
Was lying there next to him.

"I'm alive?"

"You were sick," she said,
"Something you're not used to in my world,
But you are recovering now.
I suppose it's a sign of age too,
For we've spent many years together."

"We're growing old together, aren't we,"
He continued.

"Indeed, but we still have many good years left.
Here, I'll read you something from Wordsworth
That he wrote in his later years:"

What though the radiance which was once so bright
Be now for ever taken from my sight,
Though nothing can bring back the hour
Of splendour in the grass, of glory in the flower;
We will grieve not, rather find
Strength in what remains behind.

A shade passed from between them—
A door between their worlds had opened
To let their dreams pass through.
One shooting star after another
Signaled these wishful events.

They awoke that morning from another dream,
Or perhaps they dreamt that they awoke,
On the shore where they had once discovered
The Spirit of the Earth.

They rubbed the sand from their eyes
And opened their minds to the day,
Being careful not to clear from them
The shadows of dreamy visions.

Their night-time apparitions
Had been soothing, calming,
Relaxing, real, tranquil,
Refreshing, restful, and peaceful—
Just like the water of the lake
That still slept under the morning mist.

They had camped on the shore,
In a mossy nook between some rocks,
An overhang of trees protecting them.

They couldn't see the sky,
But they could see a reflection of the sky
And its clouds in the water when the mist lifted:

A reflected bird flew in a reflected sky.
Water lilies floated in the heavenly mirror.
Orange day-lilies nearby told them
That that deep summer was upon them.

Haunting visions poured forth,
As they looked at the image of the sky in the water.
Soft winds rippled the water ever so slightly,
And blew the branches of the reflected trees.
Dreamy visions held them still sleep-eyed.
Again their worlds had met at twilight.

A lark rose from the water
And flew into nothingness.
Gossamer threads ran from rock to rock,
Seemingly attaching them to their dream world.

Was it dawn or dusk?
In the half light, it did not matter.
"Which is real and which is an illusion?"
She wondered.

"Do we sleep or do we dream?" he asked.

She answered with a poem:

Some say that gleams of a remoter world
Visit the soul in sleep, —that death is slumber,
And that its shapes the busy thoughts outnumber
Of those who wake and live —
(Shelley)

Blossoms fell from the trees,
And began to cover their feet.
When a cushion had been formed,
They sat down to prepare a breakfast
Of nuts and strawberries.

Flowers gently cascaded onto them
As their dreams took wing.

A unicorn wandered by,
Its existence fed
Only by the possibility of being.
A chimera came forth
And ate nuts and berries from their hands.
Faeries danced between the flowers,
Caught only by a believing glance.

Elves rode flying horses,
And centaurs walked proudly
Down the path near them.
These were the creatures who never were,
All living in the land that never was.

They looked into each other's eyes,
Reflecting on their thoughts.

"I'm not sure what world we're in anymore,"
She noted.

"Nor does it matter very much
Which side of the looking glass we're on,
For we are here."

"It's as if some ethereal beauty
Has descended over our thoughts,
And lent a poetic vision to us,
A shadow of some divine perfection.
It is rapt, although a little vague,
But I can sense its presence. Hear:"

—I look on high;
Has some unknown omnipotence unfurled
The veil of life and death? or do I lie
In dream, and does the mightier world of sleep
Spread far around and inaccessibly its circles?
(Shelley)

The day soon came to life,
And they saw castle builders laying stones,
Dream merchants giving away various unrealities,
Idealists realizing their ambitions,
Visionaries watching plans taking shape,
Ghosts and wraiths playing joyfully on the air,
Vapors forming and rising,
And then coalescing into forms,
Phantoms riding on the light hearted breezes,
Will-o'-the-wisps sparkling over the water,
And mirages becoming real at the slightest touch.

"I am so much enjoying our world," she said.
"Here, all things are possible;
It is an oasis untouched by oblivion and regret,
Free from contagion, debt, worry,
Care, strife, and woe."

And so they lived in the clouds,
Drifted into the Land of Nod,
Resided in Never-Land,
And made a home
In the world of make believe.

Twilight was yet to fall
And brooded awhile at the shore.
They looked at the water,
And saw therein a reflection of the sunset.
Reflected fire burned through reflected clouds.
A fish swam through the reflected sky.

She walked to the water's edge
And looked into it,
Expecting to see her reflection there,
But she was surprised and pleased
To see his instead.

"Come," she said, "look!
Come here to the shore."

He walked down to the water and looked in,
Seeing not his own reflection,
But a reflection of her.

"We have merged," he said, "we are one.
We will be strong now.
We will survive in either world."

ALL TOGETHER WRITING BECOMING

A long time ago I read all of Shelley's poems,
He being a scientific romanticist known,
Who plumbed the depths of mystery,
And too Keats and Byron, as eagerly,
They being the romantics of the earthly realm,
Along with Omar Khayyam, the Sultan's helm,
A romantic scientist who invented algebra,
As well as cherishing all of nature above Allah.

Omar was as Mr. Spock's logic,
But with the glory of life added to it,
While Shelley was more of Dr. McCoy's
Excessives of emotional romantic ploys,
But Keats and Byron were more
Of a blend, like Captain Kirk's sure
And dashing action tempered with reason—
A man for each and every season.

So, I ended up writing poems in the styles
Of Shelley's and Old Khayyàm's wiles,
The former being flowingly lyrical—
The latter twistingly epigrammatical,
Short ones at first, very precise,
But also using them as a concise
Way to whittle down entire books
To the few gems and pearls in their nooks.

So now after many educated years,
I still use them to boil down the idears.

A MEETING WITH RUMI

Welcome. What bringest thee, friend?
There, on some remoter shore of human soul
To which I helped restore life and spirit,
I learned that love was the only flame that lit
This life, for she had taught me how to give it.

What once I was has dimmed, physically,
But, I am a star, still bright in the night,
Though, when the sun rises, I disappear into her.
For no one looks for the stars when the sun is out.

No, I did not just disappear—
I am just completely soaked in her qualities.
The drop has become the ocean—
Now I drink from her spring of eternal youth.

Do we feel some memory of elsewhere?
Do we dare to look into the setting sun?
It shines through us, illuminating us.
We re-energize. We become supernovae.

What flaming forge fires all that we know?
What do we seek? We long for the TOE—
As the human mind turns to the inward sown
And thence outward as well to find its way home.

Why do we wander around in the starry dark,
In the middle of the night, as this lighted spark?
Well, if we knew the answer to our vertigo,
We would have been home some hours ago.

Where would that be—wholly home? My voice says:
I don't know—mind ever seeks. Whatever
Brought me here will have to take me home.
Or this is home and we're already there.

How do we see this home from our newer house?
Close both eyes, to see with the other eye.
Then how do we hear of it with our ears?
The blossoms drop their blessings all around.

What quenches our thirst in this life of ours?
Break the wineglass, this earthly cup of thine,
And fall toward the glassblower's breath and drink.

Why?
We are the sweet cold water as well the jar
That pours it. Plus more—we are even
That which makes the drink taste so refreshing.

Where do we go to know, climbing mountains,
The Himalayas, to find there the wise old man?
No, for a mountain is but a little piece of straw
Blown off into the sheer emptiness of the All.

What shall we feast on? The before and the afterly?
No, for we taste this minute the time of eternity.
We have wet our robes in the shallows of mirth;
Then we dive deeper, under the fathomless surf.

We're not afraid to feast
On the sweet taste
Of eternity this minute?

We dive under, even naked under,
And deeper under the fathomless surf,
Wherein the drop becomes the Ocean, too,
As the Ocean, as well, becomes the drop.

Where is the light that shines to make us so?
It was born of the many stars in that milky glow,
And so there is a light seed grain deep inside you;
You fill it up with yourself, or it dies, to embers few.

And what of her, the beloved beyond?
There is a window open across the pond.
How's that? The quiet airs mix our beings.
There's a unified field. Go forth, singing.

Out beyond the ideas of wrongdoing
And rightdoing, there's the underlying.
Go forth and then wait; you will meet her there.

And then do we see the bright light of day?
Ever this day that we sought is inside the way
Of living and dying, sunrise, sunset, and noon.
Blossom—lest the petals wither much too soon.

Did we not tire, ever walking, looking, lame?
At first, we did, yes, but then the beauty came—

The grand moment of wings grown; lifting, new.
That rhythm flies us—the music plays through.

From. . .?
'Twas fashioned even before it was.

Where have we been through all of these scenes?
Well, everywhere, and nowhere—as but in-between.
Come home! There was never the less or the prime;
And then you will know this place for the first time.

I drink the very wine that moves in me.
I freely let life's spirit play through me.
I'm its rhythm and music and live it.
Life, though rough sometimes, must be lived fully.

I spring into another level of being,
By "dying into life", so colorfully,
Like a spring flower—the energy was there
In the bulb all along, deep within.

There's a longing, between Body and Soul,
That reassures us when we go with the flow,
And tugs at us when we don't—an undertow.

The world crashes, out there,
But the flowers grow, in here.
For, we are the garden.

ENCHANTED JOURNEY

Here the markers to signal the faint path—
The Persian flask and two green fairy lamps,
Flanked by petunias, whose sums grant the math.

From green to blue her chameleon clothes change,
While she factors in Victorian pots,
Her blue-blond hair ever more rearranged.

The vase and the flagon winked straight away,
She now back to blond, sporting lighter clothes,
As she passes through the magical archway.

A bridge appears where there was none before,
On an extra green leaf morning newly born,
Spanning the gap between the river shores.

Down by its side is a mossy green log,
And just beyond, the rushing waterfalls,
And day lilies to herald summer's slog.

Here more lamps, and swans, on a forest walk,
Past an old mansion, toward sunlit uplands,
After the pure pond and its reedy stalks.

A brick-paved turn in the enchanted woods
Brings forth leafy stone gardens, in dim light,
For minty refreshments, and other goods.
f the gods
Stony steps lead upward, and there she poses,
For all the woodland's hidden eyes to see,
Flaring wide her wings, and standing tip-toed.

More falls call and so she walks, rock to rock,
Cooling off in the water's splashy spray,
On her wondrous journey that has no clock.

She lays by the side of the moving pool,
Looking closely at the swirling wavelets,
Oft catching sight of a sparkling jewel.

Next, of all things, sits a wonderland girl,
With a large hat, holding a big red egg,
On the other side of the whirling swirl.

Now with orchid wings, she sits near the stream,
Amid the moss and the caressing mists,
Lost in time, while wandering through daydreams.

Within the rocks a lime green cave was carved,
Where jewels and a golden vase were stored,
With the urn that Keats' wrote an ode upon.

There an elf's house, in a haunting face tree,
And more lanterns to guide one through the deeps,
All laid out with the words of poetry.

Here another woodsy home—a mushroom hut,
With a pumpkin and stools set for sitting.
She outstretches her arms, welcoming the night.

Another house, built into a sturdy tree,
Has two young fairies kneeling on the lawn,
Near the large stones set in the greenery.

Night falls, greeted by elemental tree-men
Guarding a massive face trunk just beyond;
Yet unworldly lights still shine in the glen.

Green and bluish magic glows light the way,
While angel sparks draw the traveler on,
Through the warm mid-summer's night, come what may.

On warm-hearted ground, under stone-cold sky,
Her carried waving lantern's light reveals
A kindred soul, set to sleep in a cup.

She finds a glowing abode to sleep in,
In sojourn with two other wanderers,
Pleasantly tired—waiting for dreams to begin.

She dreams of comets in the starry night,
From the wellspring of visions sparked so bright—
Phantasms ranging among all the sights.

Up at dawn, she runs with its orbed glow-balls,
The sun racing behind her to catch up,
She off again in her world without walls.

The morning mists and clouds diffuse the sun,
As there she stands, wondering at cattails,

Her band of blue sporting its blonding run.

The warmth surrounds her, and so she lays,
On the purple-eyed grass, near the rippling pond,
Deep in reverie of the newborn day.

She drinks the dews from the flowers before the day
Can heat and evaporate the vapors away,
Gaining this light sustenance, as fairies may.

She flies through the grasses, around a tree,
As an angel atop a mushroom point the way,
Leaving smoky swirls and vapor trails freed.

Here the deep and preternatural forest,
Its primeval splendor undisturbed,
The place where angels breathed into man's breath.

Here a stream that she will make an ocean,
Through her growing small, by quickly shrinking,
To ride the waves and crests with emotion.

Butterfly wings sail her rounded leaf boat,
That has a flattened splinter for a seat,
With a twig-vined rudder to keep afloat.

From the overflowing wellspring of joy,
Passion and desire, they drink elixir,
Deep, and wetting through and through, o'er noise.

Wanded fairies three share a wide tree stump,
In the wooded green of all greens ne'er seen,
Where winter's scene is ev'r summer trumped.

Perpetual roses spring to one's hand,
And all around, in the secret garden's bower,
Where one's romantic dreams come true as planned.

Oh, kiss me, dear, and lift me from the grass,
And twirl me up and above, in love's whirl,
Unto the heights, where most joyous times pass.

In the unknown castle, she wines and dines,
Then steps out, onto the balcony ledge,
Noting the water-falled lake and the pines.

Her odyssey continues, past a well,
On slate stones, to where giant flowers sprout,
Through deep, enchanted lands, where beauty swells.

Oh dear, there's talk of a young 'naughty',
But it wasn't that really, specifically,
But that the fae was not enough naughty!

Two lithe and barely dressed fairlings introduce
Butterflies to the blossoming flowers sown,
So that thus they help them to reproduce.

If it ev'r gets too hot, the fairies shrink,
And then find refuge under a flower's shade,
And there find droplets of dew to drink.

Under a corolla, the crystal ball
Of marble is read, to get the weather
And the future, which is why they never fall.

In the landscaped gardens, the cardinal bathes,
While all the colors mix and shift the gaze,
As a statue with a ball in the air plays.

A lantern's clue turns her purple-violet,
Expands her wings, and reverses the scene,
Erasing a few objects here and there.

The half-good, half-bad guardian says "Stop!",
That "It is the Queen's Magic Mountain Land."
Yet she relents and waves the beauty through.

A pause, while her own butterfly self sprouts,
In a flutter-by, then alights on her hand,
As they rest on a stone arch bridge o'er a stream.

Up the mountain, on a purple brick path,
As walking into thin air, with only
Grasses between one and the lands below.

The red-haired queen welcomes the traveler,
"Come; you walk the road that all must enjoy.
What do you seek of your passage, my dear?"

"I seek the here and now, nothing more."
"And so you have done well in arriving here.

We have a grand festival going fore."

Fairies dance and fly, on the mountain top,
To celebrate not anything at all,
But that of being, living, and moving—lots.

Singers range and stand atop elven huts,
On mushrooming stools, belting out the hits,
For all to hear and the dancers to follow.

The faeries bathe in a pond of potions,
Insuring their skins to be everlasting,
Pure and deep—providing beauty's notions.

Many naked swim, to better their selves purify,
In this land where timeless beings beautify,
Where ideals flourish, high in the sky.

Into the glacial lake is the special plunge,
For here is the purest water known on Earth,
The liquid that washes mortal worries away.

Evening bells ring in parties and social talk,
Lit by lanterns hanging from the branches,
Where they walk the talk and talk the walk.

An enchantress juggles the moon, in rhyme,
Sitting on a rock in the magical land,
Where lilacs and daisies grow at the same time.

At the stairs to sleep, a ballet is performed,
On the grass, with plenty of footed point,
With jazz, tap, lyrical, and other forms.

They ready for sleep in warm cave portals
That have green and leafy floors, to cushion
Their night's restings, along with flower petals.

The nightmare and her foal are both banished
To the brightest parts of night, where they all
Can but learn and grow from the wisdom of a child.

Now the aches and pains of the day can be
Swept away, on this greensward at night,
Infused by the light of eternity.

Our life traveler basks in the moonlight,
Preparing for the flight of weariness,
Through the subconscious, restorative might.

The full moon brings its light to the night,
And to the sea, on the waves at the shore,
Dimming even the watchtower's giant light.

Others head for the well of sleep and delight,
On the phosphorescent soft grasses' pillows,
Here in the fabulous land of good and right.

"I bid you rest under the stars and the moon,"
Says the Queen, "to all of you in-between,
As sleep's circles close in to let you swoon."

Dawn brightens the lake while the skies are yet purple,
And washes over one with that feeling
That a new day has just begun its fun.

They welcome the birds, who are already up,
Rising to the glow of the east, from sleep,
In the glorious half-light of twilight dawn.

She flies down the mountainside—a dream's truth,
Over ponds and grasses grown from the valley
Fertile from the rivulets sown by the mountain.

Up and over, through the misty vapors,
She wings, into the here and now of this day,
In which everything is possible.

The slopes level into meadows of bouquets,
As quickly she rambles, through rock and field,
Becoming and joining with every living thing.

The flowers roll unto a village in Utopia,
On a stream, where a young lady stands firm,
Upon a green grass platform jutting out.

And another, near an opening in tree,
With wooden planks seeming to walk by,
All under the rays of the sun's majesty.

There's a lively playground of shroom stools,
Amid rocks, plants, flowers, grasses, and trees,

For those in recess from the fairy schools.

At the base of the falls, she meets a water nymph,
With webbed feet fins and scaled legs,
But otherwise a fairy though and through.

And twins, sitting, amid scattered leaves,
Who are early autumn fairies 'leaving'.
Oh how does the summer pass so quickly by?

Within a week, the leaves flutter down more,
As a butterfly faerie sits against a tree,
In the presence of the main autumn fairy.

A broomer next appears, sweeping a patio,
With every tree about her scene being orange,
Which is an even brighter color in FairyLande.

Finally, home and heart, in her little shack.
Did I get where I wished to go? she thought.
Yes; it was the journey, not the destination.

BROTHER BILL

The CMA Leader flies his Harley Eagle,
On the road to Recovery and giving,
Teaching the gift of the Holy Spirit to others
And listening to them to pray for God's good Will.

He rides the wave for Jesus,
Surfing the ups and downs of life,
A gentle soul helping all to cope,
He himself even plagued with illness.

He's your brother, a prayer partner,
And a man of the community.

The Eagle lifts unto the sky
Upon its last flight on high,
Winging the wind aloft,
Towards Heaven's nest.

He's feeling free now;
We will always remember him.

www.ingramcontent.com/pod-product-compliance
Lightning Source LLC
Chambersburg PA
CBHW071355170526
45165CB00001B/54